GW01539764

Peter Foerthmann

Boatbuilding - Yesterday and Today

Trends, preferences and priorities

Copyright: © 2020 Peter Foerthmann
From German into English: Chris Sandison
Cover & Set: Sabine Abels | e-book-erstellung.de

Cartoons: Inga Beitz–Svechtarov
Fotos:
Peter Foerthmann: 14, 43, 47, 50, 57, 59, 60, 62, 61, 63, 75, 80, 100, 122,128
24, Edward Burnett
38, René Bornmann
52, Sebastian Groth
67, Kai Greiser
68, Jörg Jonas
94, Jimmy Cornell
102, Harry Schank
110, 128, Sybille + Christian Uehr
123, Christian Heusinger
126, Sepp Koch
154, Rainer Woehl

Publisher:
tredition GmbH
Halenreie 40-44
22359 Hamburg

978-3-347-21813-0 (Paperback)
978-3-347-21814-7 (Hardcover)
978-3-347-21815-4 (eBook)

All rights reserved. No part of this publication may be reproduced, translated, distributed, or transmitted in any form or by any means, including photocopying, recording, or other electronic or mechanical methods, without the prior written permission of the publisher and author.

Content

FOREWORD	7
BOATBUILDING	10
Materials and processes	14
Lifetime and maintenance requirements	21
Rationalisation: the key to creating value	22
Custom builds versus mass production	24
Use, intended and actual	29
Boatbuilding and the market	31
Value and resale	33
Class associations	36
The second-hand market	38
Metal yachts	43
Cost of sales	47
THE RUDDER	50
Watch out for the crunchy bits …	52
Advice: USA versus EU	54
A return to reason	57
Easy on the eye	59
THE STEERING	64
Tiller steering	67
Wheel steering	68
The Golden Globe Race	71
Treacherous waters	73
Emergency tiller	75
Drive systems	77
Shaft drives	80
Z-drives and saildrives	82
Saildrives	83
The hard facts	85

WHY "ACHILLES' HEELS"?	87
Things that go bump in the night	88
Murphy legislates	90
Soft impacts	92
TRANSOM ORNAMENTS	**94**
Windvane self-steering system	99
Time at the back	100
Local strength	102
Swim ladder	104
Gangway	106
Bathing platform	108
Davits	110
Dinghy	112
Outboard motor	114
Antennas	115
Wind generator and solar panels	116
Antenna arch	117
Bimini	120
Life raft	123
Hydrogenerator	125
The full house	126
BLUEWATER ADVICE	**129**
Thou shalt advertise	132
Principles	135
THE PILLARS OF INFORMATION PROCUREMENT	**136**
Sponsorship	138
Know-it-alls	140
The media	143
Bluewater seminars	144
The paywall	146

THE DREAM	150
Sailing: fun or serious business?	152
IN CONCLUSION …	157

FOREWORD

Few topics are as controversial among sailors as the quality of the boats we invest in to help realise *our dreams of a life afloat*. Obviously, choosing a boat ranks as one of the key decisions for any sailor: a choice that proves to have been a poor one can have serious and farreaching consequences – and is by no means straightforward (or inexpensive) to rectify.

Take on too much of a financial burden, leave yourself too much to learn or trust too much to your own handiwork and you could end up in trouble, especially if the pressganged family, left at the mercy of your orders from the bridge, also has to live with your misjudgements as immortalised in composite, aluminium or steel.

A wise choice right at the beginning can make all the difference in the world to the fun to be had under sail, so it seems astonishing that so many of the issues involved are so seldom properly discussed. Polarised debates of the *traditional = better/modern = worse* type might be very good at filling pages on internet forums (in the process bringing out the intransigence, entrenched positions, rapid loss of perspective and ruthless jump to character assassination that so often typify this platform), but they utterly fail to appreciate the complexity of the issues involved point. Rough seas ahead!

It is extraordinary really that these matters receive so little consideration in the sailing press. Are our journalists perhaps too concerned about the risk of biting the advertising hand that feeds them? Have existential imperatives led them to wash their hands of the whole notion of quality?

Certainly any criticism now tends to be limited to fairly trivial points such as headroom, berth length and interior furnishing, with the fundamental properties of a product as a seagoing vessel seldom even considered, let alone subjected to a rigorous comparative investigation.

How about some scrutiny of design trends, for example? A hot topic perhaps? The transformation of boatbuilding from a craft into an industrial process and, in particular, the implications of this change in terms of the quality and price of the product seem to me to be funda-

mental issues for our sport and yet they are hardly eating up the column inches.

Nothing sets the opinions flying like an openended debate on then versus now but there are a number of distinct aspects to consider if we are to do the subject justice:

BOATBUILDING – FROM COTTAGE INDUSTRY TO MASS PRODUCTION

Clearly sailing could never have become widely accessible as a sport without the introduction of mass production in glassreinforced plastic. The smaller the boat, the bigger the potential market open to it and hence the more units to be sold. The *Optimist* and *Laser* dinghies have been churned out in their millions and while yachts have never shifted in these numbers, their extra features – everything from heads to an engine – make them far more interesting in terms of the value to be extracted per unit.

When yacht production first began to move onto something approaching an industrial footing, the small number of manufacturers offering boats at shows in Europe could sell dozens of each model every time they exhibited. I really do mean dozens too. Back in the 1970s I would often pass the evenings during the *Interboot* boat show in Friedrichshafen playing cards with Peter Schmidt, the man behind the *Sirius* yard. One year the weeklong show brought him more than 30 orders (the celebrations left him hard pushed to stay upright, never mind cope with a hand of cards).

The Seventies and Eighties were gold rush time as sailors beat a path to the boatyard gates and demand (and sales) soared. Brian Meerloo and his *Cobramold UK* team in England built thousands of their *Leisure* yachts, for example, back when nobody seemed to worry about breathing in *styrene vapour* and extractors were unheard of.

Coincidentally it was to Brian that the founder of my company, John Adam, turned in 1968 when he needed something small and seaworthy. The *Leisure 17* Brian supplied took John West, West and West some more until, having fallen asleep and run aground, he found himself on the beach in Cuba. It was probably while cooling his heels in one of Castro's prisons that he made the decision to set up *Windpilot*: the first model had after all passed its test with flying colours (keeping watch – then as now – not being a part of its remit). John's story created a sensation at the time in Germany and all the marketing professionals in the world could not have conceived of and coordinated a better market launch for *Leisure* yachts. The brand retains a special aura to this day, with over 4,000 Leisures sold and almost all of them still in existence. But I digress ...

The number of people eager to get afloat was simply too high for traditional production methods and craftsmanship to cope. Consumers were no more inclined to wait for satisfaction then than they are now: once they felt the itch, they wanted deck beneath their feet and pronto. When it comes to impatience, children on Christmas Eve have nothing on grown men and women waiting for a new boat to arrive. Suddenly even extra costs become acceptable if they promise to bring delivery forward by a few days.

The transformation in the nature of boatbuilding precipitated by these pressures was dramatic. The process was – almost literally – turned upside down (or inside out, depending on your point of view) and ever since, boats have been built the wrong way around: today everything starts with the skin and not the bones. Henry Ford had his first assembly line up and running by 1914 but it was not until a good half a century later that the boatbuilding industry first sat up and took notice. People had other concerns in the hectic postwar years.

Originally the hull of a boat was a very complex assembly: to *keel* and *keelson, stem, sternpost, stringers, deck beams, floor timbers* and *ribs* were added planks as thick as a seaman's thumb and the whole thing was bolted, nailed and glued together, caulked and sealed repeatedly and then finally treated to several rounds of painting and polishing to create a thoroughly robust and watertight structure. All of which took some time!

A finished hull of this nature came ready to sail; there was no need for extra interior reinforcement or fittings to stiffen the structure and dissipate the working loads. To keep the weight down, racing yachts often started life with nothing below but pipe berths and were only fully furnished at a later date – in preparation, perhaps, for a second life as a cruising boat – once deemed *uncompetitive*. Whether to hit the racecourse with a crack crew or embark family and friends and concentrate on seeing and being seen was a matter for the owner: the yachts were tough enough for both lives – in fact most them still are.

In those days the cost of the hull made up an eyewatering proportion of the overall price of the boat. Boatbuilders were craftsmen, not magicians, and bending wood takes time (not that man hours were at all expensive in that era of course). Rationalising hull construction became the logical way to go – and it is here that the rot set in!

Materials and processes

Once a GRP female mould had been produced, readymade boat shells could be turned out one after the other like so many bread rolls in a process that suddenly no longer needed the expertise of skilled boatbuilders. No sooner had one hull cured than it was lifted from the mould to make way for the next.

Mass production techniques were not adopted in all areas of the manufacturing process at the same time, however, and it is characteristic of the *modern classics* that although they have a GRP hull, they were still fitted out by craftspeople: skilled boatbuilders still had a hand – and earned a crust – in their creation. Also no secret is the fact that older GRP boats tend to be considerably more solid, because their *stringers, floor timbers* and wooden interior components are all laminated into the hull. The resulting structure is very stable, especially since the stringers and bearers are bonded faceon and the interior components edgeon. The three *Hanseat* boats I have owned were typical products of this time, when sailing boats were still *seen and not heard*: perfectly quiet to sail whatever the wind and sea state and no creaking from the deck – even when the summer hoards stampeded across on their way from the far end of the raft to the bar.

Modern classics enjoy great popularity today because of the way they combine the beautiful lines and traditional craftsmanship of old with modern materials. And, at a more basic level, because of the way they allow families – even those members with a more sensitive nose – to take to the sea without the damp reek of stale air and musty socks that tends to pervade the more traditional of traditional craft. There is more to boatbuilding than right angles; indeed their very curvaceousness accounts for a big part of the appeal of those striking little ships whose lines alone set hearts pounding and bank accounts quaking. On a more practical level, the distinctive interior typical of the *modern classics* also ensures that no matter how soundly you slept, you quickly remember that you are on the boat and not in the house. Right angles simply never featured in this era, which for me explains much of its charm and elegance.

The visual language of traditional design was always emotional: function followed form and not vice versa. Today the mainstream marches to the beat of a different drummer, with aesthetic concept and product shaped by marketing diktats and design vocabulary serving only to ensure brand recognition and distinctiveness. Sometimes even a straightforward coloured stripe will suffice.

The boats of the past were undoubtedly more robust. They had a backbone and ribs, meaning they were better armed against ramming, stranding, grounding and other extreme encounters, and carried their rudder firmly mounted and safely tucked away in the *sweet spot* at the trailing edge of the keel. Modern yacht design specifies no such backbone. It has been spirited away to be replaced, here and there, with other structural members because space, increasingly, takes priority over everything else. Much the same thing has happened with cars: once upon a time they had a distinct chassis but now each component seems to be supported by nothing more than the components around it.

Interior shells and moulded parts have increasingly driven traditional boatbuilding out of interior construction too on the basis that separate units and components can be massproduced rapidly and installed (glued!) inside a bare hull equally quickly. Boats today are built in a modular fashion, which simplifies customizing and cuts costs. Moulded components can also incorporate rounded edges and corners at no extra cost to suggest at least a token element of design flair.

Today hulls are reinforced in critical areas with bulkheads, interior shells, longitudinal stringers and space frames to make sure that when the boat is lifted out, the keel comes too, that the engine doesn't slowly make its way up the boat while motoring and that the pressure transmitted from the mast above when sailing close-hauled does not leave hapless crewmembers marooned in the heads. The most extreme designs include additional metal structures to distribute the loads.

Some would have us believe that modern boats are more solid than ever, but do their claims hold water? Space has become a more compelling argument and the obvious response has been to do away with the traditional foundation of a strong hull – its frame and ribs – without putting anything obvious in its place.

Sailors and families flock to the resulting space(ous) ships like moths to a candle, for it seems the number of berths has become the measure of all things (and has a not insignificant effect on the price). A large proportion of current designs came into being with the requirements and imperatives of the charter industry to the fore too, of course, and they sell bunk space by the week!

Did anyone ever stop to consider how this wonder came to pass? Does anyone seriously spend any time thinking about how much – if any – effective loadbearing structure there is left behind that smart interior shell? Perhaps it's all a matter of perspective (after all, what we don't know might not hurt us, right?). There used to be a hull at one of the big boat shows that had been sliced in half longitudinally and I couldn't help noticing just how little substance there would be between the sailor asleep in a bunk of a heeling yacht and the fish swimming past outside. This cutaway half-hull provided the backdrop for countless Bobby Schenk seminars, so perhaps – perhaps – I wasn't the only one to wonder at the eggshell separating the wet from the dry.

It is no secret that mast, rigging, sails, engine, keel and rudder pull and push the boat in different directions and that hydraulic systems can put a bend in the hull as well as the mast. Only a stiff hull can endure such loads in the long term; once flexing sets in, serious problems are sure to follow. Signs of fatigue on the hull, keel and rudder, soon our constant companions, auger well for surveyors and the refit and restoration business. Everyone understands that the half-life of a new stripped-out racing machine is short and that, at some none-too-distant point in the future, a once proud and resolute bow will begin to acquiesce a little as the sea builds. *Life is a compromise* – and when it comes to rig tension, it can be a fine line between just right and ever so slightly too much.

We should also note that thanks to cheap materials and labour – and the fact that in those days nobody realised just how thin they could go – laminated hulls on older boats and those built in the Far East are often extremely thick. Today it's a different story altogether! The greatest secret in the increasing industrialization of modern boatbuilding is probably the sharp reduction in production times and concomitant increase in value added. Why else do international investors and the pinstriped locusts suddenly have yacht manufacturers in their sights?

I heard the story once of a sailor who, out of love for his wife and in preparation for the big retirement cruise, decided to swap his dark cavern of a steel yacht for a new – and much brighter – alternative. Keen to be, as it were, in the delivery room to witness the arrival of his new home from home, he cheerfully turned up one Monday morning at the yard appointed to realise his dreams expecting to witness the final stages of fitting out. What he saw instead was a bare hull that, like him, had only just come into the building. He was astonished – and never more so than when the finished boat was handed over right on time just a few days later.

I have seen similar myself in Les Sables d'Olonnes: the project concerned, an imposing new Privilege catamaran, was just a week away from delivery and yet there it stood in the middle of the hall, deckless, looking more like a building site than a boat. Neither I, nor the owner and his wife sitting at home in Bavaria with their sea bags packed and poised, could believe everything would be ready on schedule. But ready it was – and with time to spare.

I always find it difficult to resist comparing money invested in yachts with money spent on property (albeit houses too have their weaknesses with the wind on the nose). A newlybuilt family house – the product of countless man hours' work at a moderate level of value added for all involved – in a pleasant neighbourhood now costs about € 300,000 in Germany. A massproduced GRP yacht costing a similar amount takes a fraction of the time to create – and is then very often compared in price terms to boats whose construction involves a far greater element of craftsmanship. It is hardly surprising then that terms like *quality yacht* always seem to be applied to more expensive vessels. Obviously, *oneoffs* assembled by crafts-people are going to consume more money in the shed.

It may be interesting – perhaps shocking – to re-flect a little on the hours of work that go into pro-ducing a fin-ished yacht. The conclusions in terms of the inherent val-ue of any given boat soon be-come perfectly clear: you do indeed get what you pay for (in more ways than one).

Lifetime and maintenance requirements

It can also be interesting to look at the second-hand market and see how perceptions of quality and value change once long-term use has had a chance to expose weaknesses not apparent on the boat show floor. True quality reveals itself only in long-term use and it is no surprise the sailor's brain spends a long time quietly accumulating information before eventually reaching a verdict for or against a particular brand.

We are honoured almost every week with the launch of some or other new boat, just about all of them production GRP designs. This speaks of serious rationalisation on a scale unknown – perhaps even inconceivable – to ordinary sailors.

Rationalisation: the key to creating value

Manufacturers have found ways to rationalize their business in all kinds of areas: in design, which is entrusted to powerful computers running sophisticated software; in mould making, in which robotic systems capable of five-axis CNC machining do all the work without human involvement (with no lunch break and no expectation of being allowed home for the night); in the construction of the bare hull, which is produced using prepreg materials and vacuum bagging; with interior shells, which make it possible to reinforce and partition the hull simultaneously; with interior fixtures, which are produced as modules on a separate line and can then be installed into boats as required; and in the production of the deck, which has the fittings installed prior to being dropped onto the fitted-out hull to seal the finished yacht like the lid on a jar.

That such rapid changes of model are even possible is, probably without exception, down to the enormous rationalization and acceleration of all processes. Now, even small production runs can be profitable or pay off in the end. Nevertheless even the biggest yards with global marketing and a lively charter supply business still achieve only modest output when compared with the producers of other high-priced goods, which may well explain the breathless pace of competition: production yachts have to be introduced, promoted and phased out again faster than cars just to maintain the pressure on the competition. A design lasts just a few hundred units before the focus moves on and the next contender *sets sail*. In a saturated market, however, there is only limited scope to keep sales volumes rising. We do not (yet) have a scrappage programme for old yachts! Callous as it may sound, a lively hurricane season can do wonders for business if it leaves the charter companies and their insurers racing to replace decimated fleets with the latest models off the line.

Custom builds versus mass production

Building a metal yacht remains a time-consuming business and the preserve mainly of one-offs. A high-quality 45-foot custom boat in aluminium might take around 5000 hours to build; a GRP equivalent of the same size would almost certainly be ready to hit the water in a fraction of this time. The GRP version would surely be significantly cheaper too, so it wins hands down – unless, perhaps, other factors like seaworthiness, safety, reliability, suitability for bluewater sailing, performance in use, value-holding ability and ease of resale also come into the equation.

This is a subject as controversial as it is complex for sensitive boat owners: a sailor's position on the matter is immediately evident from what he or she chooses to sail and it is only natural – a question of self-respect, no less –

for us to want to explain and defend our views. One could write heavy books on the subject of these deliberations – in fact various authors representing various perspectives have already done just that. My take on these complicated issues is just that: my take. My expertise lies in self-steering and in helping yachtsman and yachtswomen gain a little respite from the rudder at sea. I make no claim to have eaten the tree of knowledge leaves, bark, beetles and all! The fact that after several decades in the business and dozens of boats of my own, I find my conversations with a global sailing community increasingly focusing on boats as such rather than ‚just' the choice of transom ornament, has helped to keep my job stimulating and has led me in the process to develop a particular point of view. But that is all it is: my point of view.

I believe proper consideration of these ideas based on a rigorous investigation of whether modern is always better could trigger – or at least sow the seeds of – a seismic shift in the way some sailors think about boats. Trying to phrase the arguments involved in simple black and white terms serves no purpose in the same way that no one material can always be the right choice for every boat: just as GRP has its osmosis, aluminium its electrolysis and welded steel its rust, so the plank has its worm. While we have to make a decision, we understand that (just as in real life) compromise is the name of the game. Some compromises are better than others, however, and what better way to start an exploration of their relative advantages and disadvantages than with the observations of a certain grey-haired character who, though his age might suggest a rather conservative angle, derives his opinions directly from experience and careful reasoning rather than dogma and stubbornness?

Knowledge is a valuable commodity and one it's hard to have too much of when looking for a new home on the seas. Although the ideas discussed here relate first and foremost to the yacht intended for longer voyages, it cannot do any harm to reflect on them and how they might apply whatever you have in mind. Sailors are dreamers and their dreams – irrespective of whether they ever come to fruition – almost always revolve around open water and distant islands, so it makes sense to consider how the picture may have changed over the years in terms of the quality, robustness and price of the great proliferation of production yachts in which we are now invited to trust. Lucky indeed the sailor who manages to plot a smooth passage through these treacherous waters and find a boat that lives up to expectations even after the trials of long-term use.

Owing to the changes in boatbuilding methods, industrialisation process and value-added considerations in large-scale GRP production discussed above, today's production GRP yachts are far removed from the traditional understanding of what a hull should be and how it acquires its strength. Admittedly traditional structures are not necessarily the only solution, but the sense of security their solidity conveys is certainly a good feeling to have.

I believe that boats were built better in the past essentially because the methods then employed produced a *more robust* vessel. And in those days hull, keel and rudder formed *a single strong unit*, which is always a plus in my eyes. Expedience and cost considerations dictate that typical modern production GRP hulls, in contrast, are manufactured as separate shell and keel modules that are simply bolted together when the time comes. The implications of the shell approach for bonds, joints, seaworthiness and endurance are widely understood. It is no accident that modern underwater shapes (be they trapezoidal, rounded or chined) always include bolton extremities rather than the traditional combination – guarantor of a smooth ride through waves, good seakeeping and relative peace and quiet down below even in angry seas – of a V-form forefoot and S-shaped frames further aft.

Today *speed* often seems to be the only element of performance that counts. The gung-ho yachtsman can imagine nothing worse than being passed to lee by a bathtub and inherent strength is the asset many seem willing to sacrifice – at times apparently without even beginning to consider the consequences – on the altar of vanity. Why else would it be that more and more offshore events, even those organised primarily for flotilla sailors, are making it mandatory for participants to carry an emergency rudder in order to spare the rescue helicopter the need to launch every time a steering cable jumps off a block?

A quick glance at the spacing of the rudder bearings and the leverage that can be exerted on them in most modern designs should provide more than enough material for seafaring nightmares. A spade rudder is certainly efficient in form and effect, but it also makes a tremendous lever and experience suggests the force applied at the bottom at times exceeds the strength of the mounting structure at the top. Not everyone can walk on water and those not thus blessed might do well to avoid knowingly putting ourselves in this situation rather than just trusting the sat phone, EPIRB and shortwave radio to buy us a second bite at the cherry. Bad things happened to rudders even before the orcas came looking for revenge …

One Kiwi professional delivery skipper apparently *wore out* several balanced rudders while taking his German-built mass-produced 37-footer *Kerikeri* back to the Land of the Long White Cloud.

The stories of problems with keels and rudders that fill an increasing proportion of the nautical press seldom make headlines outside of the sailing world, not just because the sailors affected often have the wit to find their own way home without any drama but also because incidents like this are so common that they no longer qualify as news. *Yachting World* recently reported that about ten percent of the ARC fleet suffered rudder problems of some description. Again, this was before the orcas went rogue.

Use, intended and actual

Performance (insofar as it is confused with speed) has another dark side of which anyone who values the chance to sleep in peace at sea ought to be aware. The number of sailors quickly and quietly trading in their fast rides for better allaround off-shore performers is rising rapidly. One family I know of couldn't sell their performance yacht fast enough on reaching the Caribbean after measuring it against the Atlantic for the first time. The decision cost them a huge amount in monetary terms but the 40-year-old classic they transferred to took them safely and happily the rest of the way around the world. Others have been less fortunate.

Swiss circumnavigator Thomas Jucker once observed: „Sailing to New Zealand and around South Africa in a lightweight boat is actually not particularly difficult – but it isn't much fun either." Jucker currently sails a *Bristol Channel* cutter. Double-acts Kicki Ericsson/Thies Matzen and Lin/Larry Pardey have covered hundreds of thousands of miles criss-crossing the planet aboard their diminutive wooden boats over a period of decades without once having cause to question their vessel's seaworthiness. All just anecdotes? Of course, but there are plenty more where they came from and they tell a consistent story – a story even ordinary sailors who do not spend their whole life at sea would do well to heed if they really want to be able to trust in their boat.

The open ocean, of course, has no monopoly on uncomfortable conditions: the combination of strong winds, powerful adverse currents and prominent landforms can put yachts to the test even close to home.

When every week confronts us with new models, glowing test reports and more slick marketing to match and when each successive development seems intended only to make us forget what has gone before, it takes a stoical composure to read between the lines and seek out the real purpose of this spiral of gloss. Clearly new does not automatically equal better. Older boats still making their way under sail have proved they can last the distance: novelty in and of itself is no reason suddenly to consign them to the yacht cemetery. Again, we do not yet (thankfully) have a scrappage programme for boats.

Boatbuilding and the market

Once more into the breach! The only way to be sure a boat is really robust and well-built enough for long-term use is to put it to long-term use – or break it trying … However the typical year for most seafarers yet to shed the shackles of the rat race amounts in total to no more than a few weeks under sail in a good year and less than a hundred hours on the engine in a bad year: most boats spend most of their time turning silent, solitary circles around a mooring buoy or staring at an empty car park in the vain hope that today, at long last, the skipper will finally put in an appearance and let them off the leash.

I wouldn't want to put an exact figure on the number of nautical miles required to reach a verdict, but I do think considering the relative merits of different boats in long-term use is a good way of establishing the criteria that really count in practice. It is self-evident that yachts used for bluewater voyaging have to withstand far greater loads than fair-weather weekenders. Strongly tidal waters and the higher latitudes also have their challenges I concede, but we cannot penetrate to the heart of the matter without considering proper long-term offshore exposure (which happens to be my pre-ferred terrain).

Value and resale

A design's ability to hold its value over time is a key indicator here: any owner whose boat is known to have come up short in the moment of truth can expect to see its resale value evaporate like sea spray on a hot day in the Canaries.

Boats become permanently unsalable if the nautical community gets wind of any serious weaknesses. Obviously the yachts we turn to as the vehicle of our sailing fantasies in generally show far more endurance than their crew. When conditions deteriorate, the great majority of sailors head straight for the nearest port, teeth clenched, without ever giving the storm sails a chance to shine. The more unpleasant memories tend to fade quickly, especially away from the water, and soon enough bad times have been reborn as great stories and it's time to get back afloat.

This, of course, is all part of the charm of our sport, but it does nothing to alter the fact that most boats are very seldom really put to the test. While this may be good for the manufacturers (nobody likes complaints), it does increase the risk of sailors only discovering the true limitations – and absolute limits – of their vessel once it is too late to find an easy way out.

We could of course turn a blind eye to stories of trouble at sea, keep our thoughts within our comfort zone and specifically avoid learning just what can happen in extreme conditions. This approach, I would suggest, is likely to be more popular among those who have reason to fear their own judgment: it cannot be a nice feeling to buy a boat for one set of reasons and then fall out of love with it on other, far more compelling grounds.

The matter of hull strength we have already done to death and there is nothing to be gained be revisiting it. But what about keel, rudder and engine? A saildrive, for example, which leaves the crew parted from the fishes by just a few millimetres of gasket, can prove the Achilles' heel of an otherwise robust yacht. Saildrives have insidiously worked their way aboard all kinds of designs now thanks to the fact that they are much quicker for the yards to install, but they remain a matter of concern among insurers and have proven by and large to be a retrograde step in the long run for us sailors.

Here too, the new 'solution' fails to measure up to the older ideas it has replaced: commercial shaft drive systems always include a thrust block to pass on the thrust from the shaft, but force transmission in a saildrive system is inherently unfavourable because of the way thrust generated in the basement has to be tamed and transferred a whole level higher up in the structure. This is quite different to the situation with a conventional drive, which transmits its propulsive force to the backbone of the vessel in a straight line and, of course, enjoys a protected location tucked in at the trailing edge of the keel.

The difference in terms of the work involved in installation and alignment between a conventional shaft system and a saildrive amounts to many hours – too many hours for the major production boatbuilders to ignore. For sailors, however, the case is not so clear cut: while a leaking stern tube is certainly annoying, it is nothing compared to some of the problems encountered with a saildrive.

Class associations

Class association publications and resources can be an excellent point of reference when trying to assess the long-term qualities of production yachts, as they describe known weaknesses and solutions and can provide useful tips to help owners make the most of their investment. Well-known *owners' associations*, some of which enjoy almost legendary status among owners and future owners, include:

Baltic, Bandholm, Breehorn, Bowman, Boreal, Bristol, Camper & Nicholson, Cape Dory, Cheoy Lee, Contessa, Contest, Corbin, Columbia, Cumulant, Garcia, Hans Christian, HR, Hurley, Island Packet, Koopmans, Malu, Meta, Morgan, Motiva, Moody, Najad, Pacific Seacraft, Passport,

Pearson, Peterson, Ovni, Rival, Sadler, Southern Cross, Swan, Tartan, Tayana, Trintella, Victoir, Vancouver, Variant, Van de Stadt, Westsail and Westerly.

Just about every reputable brand operates an *owners' association* (though some do find the temptation to use it mainly as a marketing platform for new models hard to resist).

The *Contessa 32* occupies a special niche in the used boat market thanks to its exceptional build quality. Take an example with a lifetime of use (30 to 40 years in many cases) already behind it, send it for a *full refit* with the original builder and you have a boat as good as new ready for another life-time of adventures. Yachts like this hold their value and represent a good investment for their owners. The same cannot be said at the moment for many of the more common modern production yachts, whose resale prices are seriously compromised by the sheer volume of options available in the second-hand market.

The second-hand market

As elsewhere, market activity and prices depend on the balance of supply and demand, with excess demand pushing up prices and enabling sellers of the most coveted models to walk that little bit taller on the way to the bank. If there are more sellers than buyers (say, for example, the market becomes flooded with retired charter yachts), however, boats will only sell, if at all, at a substantial discount.

A few minutes in the company of *Google* and the usual portals can give a pretty good idea of the market value of used yachts (although it should not be forgotten that there can be a big gap between asking price and actual selling price and selling prices hardly ever become public knowledge). That today's market is a difficult place even for high-quality yachts from respected yards is evident from the way so many pre-owned examples take such a long time to find a new home. The classifieds pages of leading sailing magazines contain innumerable repeat ads, with boats commonly taking several years to sell despite being advertised with good photos and plenty of detail. Brokers have no magic power to speed up the process either – in fact there is often little they can do other than add yachts to their own listings and advertisements.

And yet designs from certain yards still seem to sell with ease. Second-hand *Koopmans* boats, for example, regularly command excellent prices – irrespective of hull material – and are commonly snapped up almost as soon as they become available. *Ovni, Garcia, Bestevaer, van de Stadt, Hutting* and other aluminium yachts also often seem to change hands quickly with little fanfare at a price strong enough to spare the seller's blushes.

The stand-out performer among the GRP builders meanwhile is *Breehorn*, whose *Koopmans-designed* boats are purchased almost exclusively by private buyers and are in most cases used for long-distance and offshore sailing. Their seakeeping, construction and robustness have elevated *Koopmans* yachts in the European market to a position similar to that once occupied by *Sparkman & Stephens*.

A good reputation, which boosts resale value immeasurably, can only be earned through satisfied customers. *Word of mouth* recommendations are becoming increasingly important for manufacturers today: full-page glossy adverts carry little weight if the people who actually sail the boats featured are left unimpressed.

Without wishing to mourn unduly the passing of traditional boatbuilding skills or to brush aside several decades' worth of developments in yacht design, I believe we can safely agree on the following: modern industrial production methods have greatly reduced the amount of effort, as measured in hours of work, involved in building a sailing boat (manufacturers would otherwise have little interest in mass-produced yachts – the scale of the potential value added is what sets the big players salivating). It seems rather curious then that while the automotive industry, for example, generally makes no secret of the number of hours of work required to manufacture different cars, the amount of time taken to build a boat never gets an airing in public. We have to content ourselves with speculation – and we speculate that it probably does not take very long at all.

Also striking is the fact that while modern production yachts progress through the yard very much faster than their predecessors, the price we pay for them has not fallen at the same rapid rate. Today's volume manufacturers manage to maintain extensive ranges and unveil new models frequently despite selling but a modest number of each design. We can deduce from this that, thanks to modern moulding techniques, they do not have to shift many units of a new product before moving into profit. How else could they carry on launching new products at the current breathless pace?

With the price of a new boat now broadly comparable with that of a fairly serious piece of real estate, I do not think it unreasonable to raise the question of just how well today's craft hold their value. While marketing can affect this to an extent, sailors are very well attuned to more concrete factors – notably the quality of their floating beauties as verified in long-term use and, of course, simple supply and demand – and it tends to be these that determine the price for which pre-loved (or not) boats eventually change hands.

Some would apparently like us to believe a yacht has a maximum life expectancy after which the scrap heap beckons, but as sailors our heart, our head and our experience all tell us otherwise. Consider what happens in other areas. When it comes to a house, for example, the principles we apply are quite different. We more or less take it as gospel that in the long term, the price of a home in bricks and mortar moves in one direction only: upwards. Cars, in contrast, begin losing value and integrity to depreciation and rust from the moment they leave the showroom and only the passing of the years reveals which models have the wherewithal to bounce back as classic collector's items and which get melted down ready to do it all again in some other form.

How the value of our floating assets evolves in practice largely comes down to build quality, but it's hard to imagine any sailor accepting that his or her boat would be *sailed-out*, as it were, after a certain number of years and then being prepared to accept any price on the grounds that anything is better than nothing for an *end-of-life* product. Sailors have a highly refined sense of value when it comes to their boats. However finding a potential buyer with a similar view of worth is proving especially difficult for many sellers at the moment and while those of stout heart can usually manage a polite refusal, responding to a perceived derisory offer for a boat in which you have invested so much of yourself can be a painful business. Such situations are handled more dispassionately in the USA, where sellers seem able to accept or refuse offers freely without any sense of personal insult – emotions play no role here and tears are seldom shed.

But back to the matter in hand: the better its quality (which equates more or less to how well it is built), the better a yacht will hold its value. The venerable old-timers already enjoy a strong market and the *new classics* are following a similar path. Mass-produced GRP yachts, in contrast, are available in huge numbers second hand, in most cases at broadly similar prices, and selling has become increasingly difficult as the market slows and buyers can hardly be tempted at all without deep, deep price cuts. For months my head has been absorbing scarcely believable facts about boat sales and the prices actually paid for thoroughly respectable yachts. These are good times indeed for sailors hunting the boat that best meets their personal needs.

Metal yachts

The market for aluminium yachts, unsurprisingly, seems to operate according to a completely different set of rules. Sailors of all persuasions understand that aluminium hulls are in a completely different league when it comes to build strength and durability. They are just inherently more robust in all respects than production GRP boats. The insurers' photo archives provide a compelling insight into the toughness of boats that fall foul of extreme weather, get too friendly with another boat or otherwise overindulge in off-piste activities.

Almost all of the metal yachts built in Germany come from small-scale builders with little or no production automation, so hull costs alone tend towards the eye-watering and the price of a finished boat is seldom less than staggering. Expensive *one-offs* and superyachts aside, Germany is quite insignificant internationally as a producer of aluminium yachts; in fact the only German company to have achieved significant numbers is *Reinke*, whose hard chine designs can also be built by fearless DIY-ers (just how many projects there are languishing half-finished at the end of gardens all over Germany after would-be boat-owners found the work harder than anticipated, the costs more than they could stomach or the looming prospect of divorce too unattractive to contemplate nobody knows, but it's likely to be more than a few).

Although Germany has several manufacturers capable of holding their own internationally in the production GRP boats segment, it has seen little if any significant innovation in the production of aluminium hulls. We need look no further afield than Holland though for a good example of what can be achieved. The Dutch have quite a number of yards that have transferred the production methods learned in shipbuilding to yacht construction: computer systems control the entire process, from lofting whole sections to cutting out the various components (using plasma cutters) and shaping and finishing them, which makes it possible to produce a finished hull with minimal tolerances and very smooth lines. The result is yachts with a perfect surface finish that need no more than sandblasting; filling and painting are seldom required.

The traditional frame designs and spacings used in aluminium hull construction of course make it enormously strong too. Even grounding, encounters with ice and other sleepless-night-inducing impact scenarios need not necessarily be a problem for a well-found aluminium yacht. The fact that hulls built in seawater-resistant 5083 grade aluminium do not even need painting means they are ideally prepared for the knocks and dings of everyday use – and further underlines just how different they are to modern eggshell-thin GRP equivalents, which have little in the way of load-bearing structure to repel the advances of an uninvited rock or reef.

Aluminium even has advantages over steel, as Jimmy Cornell, who has been around the world with both, neatly summarises: on his steel boat, the paintbrush used for touching up after the bumps and scrapes of harbour life seldom had a chance to dry, whereas errant helms careless enough to make contact with his aluminium hull received nothing but a smile and *"Have a nice day!"* After ten years and something like 80,000 nautical miles of sailing, his *Ovni 430* still looked fresh as the dew and eager to set out adventuring again with her new owner.

Anyone who has witnessed a production *Allures, Garcia, Boreal* or *Ovni* taking shape at the yard in France will understand very well how they have come to be so popular. Demand and lead times are enormous and the second-hand market is stable, almost as if we had had nothing but economic good times all the way. France has always produced a large number of aluminium yachts, in part perhaps because Eric Tabarly adopted the material early on for his *Pen Duicks*, but also because French yachtsmen and -women tend to sail more, harder and in tougher conditions than their counterparts here in Germany. Yachts bearing the *Allures, Garcia, Boreal* and *Ovni* brands have a global reputation and are sought-after by discerning customers all over the world.

German bluewater veterans Astrid and Wilhelm Greiff completed their 1992 circumnavigation with a French *Via 42* called *Octopus*, a boat whose builder subsequently ran into financial difficulties as a result, I suspect, of its obsessive pursuit of perfection. The *Via* centreboarder, with its skeg- and keel-mounted rudder, is today a much-coveted rarity. *Octopus* is still chasing the far horizon, currently under the name *Mirabelle*.

Chatam, Chassiron, Dalu, Damien, Garcia, Levrier De Mer, Madeira, Maracuja, Meta, Romanee, Trireme, Trismus, Trisbal, Reve Tropique, Via: largely unheard of in the wider world, all of these yachts – some available in steel only, many in steel or aluminium and some in just aluminium – offer great stability, have an extensive history of bluewater voyaging, are lively and long-lived and find them-selves much in demand.

France, remember, is different. A true nation of seafarers, the French, seadogs and landlubbers alike, revel in the record-breaking exploits of their sailors and turn out in huge numbers to support and honour the country's nautical heroes with a passion reserved in most other countries only for football.

Cost of sales

One of the peculiar features of the aluminium boatbuilding scene is that aluminium boats are wholly absent from many boat shows (I suppose there is just no incentive for yards to attend when the order book is bulging most of the time anyway). While this alone speaks volumes, it is also probably worth mentioning again that the cost of attending boat shows, like other marketing and sales costs, passes straight into the purchase price. Hence a manufacturer that can market its yachts effectively without having to stump up for boat shows can put more of its revenue into delivering a high-quality product. Painful it may be to admit, but logically it makes perfect sense.

Nordsee yachts, all of them built by *Dübbel & Jesse* on the island of Norderney on the German North Sea coast, enjoy legendary status in Germany and are highly prized in the second-hand market. Few ever come up for sale, however: once acquired, they tend to become family heirlooms, with successive generations of sailors appreciating their outstanding quality and building a firm conviction that after this, any other boat available would be a backwards step. Unfortunately the production of Nordsee yachts ended with the tragic death of one of the yard's proprietors, Herr Dübbel. Nordsee boats have an impressive bluewater record too: not only did multiple circumnavigator Wilfried Erdmann survive his many adventures unscathed, for example, but *Asma/Toanui/Gjoa/Moli* – the same yacht under four different names – is still at it and recently chalked up its eighth lap with current owner Randall Reeves.

Benjamins in Emden, close to the Dutch border, is probably one of the few German yards still building aluminium yachts professionally in any significant number.

Skorpion yachts from the *Feltz* yard in Hamburg also have a solid reputation in the bluewater community. Feltz used to work in both aluminium and steel but after a change of ownership a few years back it now only accepts orders from public authorities and the commercial shipping sector.

The way that aluminium and steel boatbuilding has fallen out of favour in Germany cannot be explained by any lack of demand, but it certainly does take entrepreneurial vision to guide the development of a yard in such a way as to permit the production of attractive yachts from good designers using the latest construction methods at a price customers will be willing to pay.

The Dutch and the French have a thriving and technologically advanced boatbuilding industry capable of fabricating eye-catching vessels at a reasonable price, so it is no wonder we immediately look to these two countries whenever the conversation turns to metal yachts.

THE RUDDER

Appearance and reality

A tour of European trade fairs and marinas reveals which marketing campaigns and promotions have gained the most traction and highlights the latest trends and developments to have been packaged up, badged (invariably) as important improvements and sold on to the sailing public. Most producers bringing new products to market aim to square the circle of performance, utility value and living space, which may well be OK for the ordinary sailor who has no real intention of venturing beyond the reach of the maritime emergency services. Boats that are going to visit the high seas, on the other hand, need to be judged against a quite different set of criteria in my opinion. The design and strength of the appendages, for example, are surely of central importance, so why does they receive so little attention when it comes to marketing and promotion?

Watch out for the crunchy bits …

Insurance company records on the frequency and causes of claims leave no doubt that rudder design and construction have become the Achilles' heel of modern yachts. Part of the reason for this unfortunate state of affairs is that so many of the European yacht manufacturers buy in industrially manufactured rudders from external suppliers, presumably for cost reasons. Well-made as they are (see *Jefa*, for example), these rudders are by their nature installed with just two bearings – two bearings that can be really quite close together on some of the hull designs with which they are used. A skeg is often omitted, which improves the effectiveness of the rudder and reduces costs.

Anyone thinking of heading offshore with a boat sporting a rudder of this description may wish to consider investing in a guardian angel to keep it clear of anything solid hanging around close to the surface. Alternatively, it will probably not be too long now before the *Oscar* collision warning system being used for the first time on the *IMOCA* class yachts in the latest *Vendée Globe race* starts to appear on mass-produced boats as well. That said, I'm not convinced the system will be tremendously useful for cruising sailors because it obviously cannot detect objects that are completely submerged even if they are still close to the surface.

The robustness of appendages is a matter any sailor planning a big trip needs to start thinking about right at the beginning of the process (and certainly before buying a boat for the job). I have a personal interest in this sort of thing, as it happens. I am always pleased to offer sailors my advice when asked, but the pleasure is that much greater if they ask me while there is still time for my words to make a difference. If the people concerned have already put their money into (what I consider to be) an unsuitable craft, I can but politely wish them good luck with their new purchase (and hope they have that guardian angel standing by).

Advice: USA versus EU

The job of administering advice and recommendations in the US market lies in the hands of the countless bluewater sailors and sometime authors who record and publish their own experiences, very often gained aboard robust long keelers, through their own small publishing houses. The typical bluewater yacht in the US follows traditional lines, with the main rudder hung on the back of the keel or a substantial skeg. Long-distance cruisers in the US seem to have less enthusiasm for modern performance yachts.

The situation in the European market is very different. On this side of the Atlantic we are quite used to seeing in print so-called test reports that clearly amount to little more than advertorials. These puff-piece slices of favourable press cobbled together in exchange for a full-page advert or two are prone to praising much while scrutinising little – and construction and practicality at sea are two of the crucial issues often overlooked. These reports are continuously referenced by manufacturer and publisher alike and can even be found years later being offered for sale as a download. There exists a clear and present danger, in other words, that the print media sources used by sailors to help with their choice of steed are little more than just another cog in the marketing machine.

The differences between the USA and Europe in terms of the types of boat used and favoured for bluewater activities are huge. Why? It is difficult to look any further in pursuit of an answer than the fact that one region relies on word of mouth while the other relies on marketing. It seems logical that the personal experience of sailors would carry more weight than marketing measures and test reports of uncertain provenance.

I know from personal experience that products can successfully find a market without so much as a mention in the media given a critical mass of word-of-mouth recommendations.

Books and bluewater seminars can provide further food for thought for sailors with purchasing decisions to make, but only if they tackle the uncomfortable truths head on (which, as I address in my observations on the monetarisation of these seminars, might not always be as straightforward as we would like to think).

The advice now offered may well extend to include the differences and key features of steel and aluminium hulls, but there is – in my opinion – still too little attention paid to the hull appendages and to their suitability for use away from coastal waters. This applies in particular to GRP designs, some of which are produced in (relatively) vast numbers. Flotilla organisers' safety tests and mandatory equipment lists are sensible enough in themselves (ensuring each boat has an emergency tiller that actually works is certainly a big step in the right direction), but they do nothing to address the risks inherent in the basic design of these boats and the worrying implications for the offshore sailor.

My object here is to highlight the design features that really matter on boats destined for blue water – features that can become critical for survival in some circumstances.

A return to reason

The evidence in my segment of the market has been building up for some years: tried-and-tested hull forms are making a comeback. Yachts with a long keel remain popular because they apparently just make sense to a large part of the prudent bluewater public.

My list of the designs particularly often fitted with a *Windpilot* includes:
- 432 Hallberg-Rassys of all sizes
- 68 Monsun 31s
- 82 Contessa 26s, 32s and 35s
- 48 Vega 27s
- 17 Tradewind 35s
- 16 Rival 32s and 35s
- 20 Rustler 36s
- 17 Pacific Seacraft
- 16 OE 32s

- 22 Albergs of all sizes
- 17 Allegros of all sizes
- 15 Breewijd 31s
- 15 Bristols
- 42 Breehorns of all sizes
- 18 Endurance 35s and 37s
- 63 Nordic Folkboats, IF 26s and Marieholm 28s
- 14 Island Packet 35s, 37s and 45s
- 18 Laurin 28s and 32s
- 34 Nicholson 31s, 32s and 35s
- 13 Passport 40s, 42s and 45s
- 105 Colin Archers of all sizes
- 22 Tayana 37s and 52s
- 14 Westsail 32s, 42s and 50s

This is just a quick extract on the fly from my database, which covers virtually all of the series-produced yacht designs for which I have manufactured and supplied systems over the last 40 years. The actual figures, however, cover only the last 20 years.

Easy on the eye

This article by a US based owner who has become very familiar indeed with the rudder bearings on his *HR Monsun 31* in recent years gives a good idea of just how solidly the rudder is attached on such a boat. I can imagine a few owners of modern performance cruisers with a balanced spade rudder blanching at the very sight of these pictures!

I appear to have lost all shame when it comes to the never-ending subject of *rudders*. Give me the merest hint of a chance and I'm only too happy to hold forth yet again on what the ideal rudder looks like (ideal for long trips on the open sea, that is).

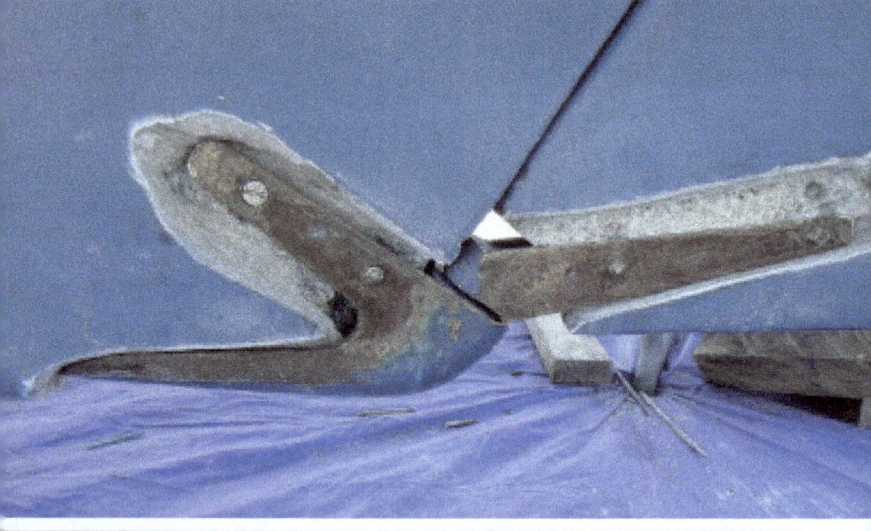

I know how sailors choose to spend their money and while I appreciate that some do eventually appear to take the lessons of their *homework* to heart or fall into line with the exhortations of those experienced speakers on the event circuit, my day-to-day experience in the business tells me that there is still too little attention paid to the rudder, its design and construction and its characteristics. It is like being trapped in an endless loop, somewhat akin, I can imagine, to having teeth pulled while fully conscious.

THE STEERING

Is overconfidence the order of the day? Have we lost our respect for the forces of nature? Naively succumbed to the marketers' forked tongue? Is the new really always better than the old? Are today's boats objectively better than yesterday's? Deep down inside, where the flame of *personal experience* bravely resists the rushing winds of sales-speak and the blustery but hollow protestations of the self-proclaimed oracle behind the keyboard, we know the answer to these questions.

What-if, what-if, what-if? What if we really delved down into all the what-ifs? Less sleep, for one thing! But we sleep easy after all: who has the time to think about the darker side of what may be? Lives are busy and the clock runs faster with every passing year. Sailing brings a much more relaxed pace to life, of course, but with the boat, the sea and the wind competing for our attention we are highly unlikely to give any thought to the potential negatives of our sport – unless fate happens to bring us face to face with them.

Or is our failure to consider what could go wrong actually a conscious decision? Do we in fact hear that inner voice of caution, those unwelcome words of wisdom and actively suppress them? Perhaps we should gird our loins and give our intuition free rein after all. Perhaps it might be worth making time to think about what can go wrong and what we might like to do about it?

I would like to shine a light into these dark places for a moment, to dust off the old mantras and, at the risk of raising the collective blood pressure, to see whether we would not in fact do well to remember the lessons our seafaring ancestors have bequeathed us.

The rudder is the soul of the ship. Loose or damage it and reaching any safe harbour in any kind of shape will be a triumph. So wholly reliant are we on this one component that the old-time designers would never have dared to suggest anything even half so vulnerable as a spade or balanced rudder. They liked their rudders tucked away at the aft end of a long keel, where they could be robustly attached along their entire length and were thoroughly protected by the keel itself, which would reliably bulldoze anything solid out of the way.

However many reasons we might hear to move beyond the old rules and cast off the yoke of past principles in favour of the modern and the new, *safety at sea* ultimately still trumps every other concern. All the sharp theory, clever words and *sofa-bound wisdom* in the world count for nothing when fate strikes, fear breaks out and physics unshackled takes control.

Tiller steering

My predilection for traditional types of boat has its roots in their particular suitability for serious off-shore sailing. My special interest of course is in sparing the crew the torment of endless helming, which probably explains why yachts with tiller steering and the rudder solidly hung on the keel or a skeg are for me the brightest stars in the blue-water firmament.

There are 40-tonne *Colin Archer* designs with tiller steering that are as light as can be on the helm – a delight to steer for even the most delicate of drivers. The Vikings (not the most delicate of drivers, I think we can assume) steered very substantial boats by hand using solid wooden tillers, along the way giving English the word starboard, among others. More recently, the *Golden Globe Race 2018* was open exclusively to long-keelers with a keel-hung rudder (all designed before 1988).

Wheel steering

I have always wondered how many sailors prefer wheel steering simply because of the photos: how much more imposing one must look standing at the helm square-on to the world rather than perched on the side-deck clutching a tiller. There are of course also plenty of good reasons one might prefer a wheel. Assuming though that the rudder is properly attached to the keel or skeg, the mechanical wheel steering system is the weak link in the chain: the rudder is operated indirectly and the additional components required as compared with tiller steering (cables and blocks, push rods/gears or hydraulic lines) all represent *potential failure points*.

Modern cruising boats of the type also thought up, built, sailed and promoted for bluewater use are, however, aligned with a different set of priorities. Many yards buy in mass-produced rudder blades (complete with shaft) and then fit (weld/bond/laminate) them into the hull. Not least because of their carefully profiled and balanced design, these rudders are installed well aft, far from the safety of the keel, and rely solely on their shaft and its mounts and bearings in the hull for survival. Obviously specialist companies with their specialist expertise are able to produce better quality rudder systems at a better price (with more profit for the boatbuilder), but a measure of trust is nevertheless required.

The vertical separation between the two rudder bearings and the leverage exerted by the rudder blade below raise questions about the ultimate robustness of this approach and such rudder systems certainly cannot be considered the equal of older designs when it comes to safety. The effectiveness of the rudder in performance terms has clearly come to take precedence over strength of mounting and attachment, at least in the context of the ideal solutions described above.

Not surprisingly, almost all of the large-scale boatbuilders now choose to procure saildrive units and complete rudder and steering systems from specialist companies rather than manufacture their own solutions. Presumably the financial benefits are just too significant to ignore. The risks and side effects of this decision are left to the sailor to bear: a spot of water ingress around the saildrive seal or a little "love tap" between a balanced rudder and anything solid will generally suffice to bring the downside of the modern approach into sharp relief.

Provided that nothing gets in the way, containers, fishing nets and other partially submerged solids keep clear, no uncharted shallows or reefs pop up at the wrong moment, the wind and the weather play ball, all on-board systems remain fully functional and the skipper doesn't leave anything to luck, thousands of joyful sailors will continue to proclaim at top decibel on reaching their destination, "We were fine. It's all a lot of fuss about nothing!"

Anyone interested in building a more accurate picture of what can and does go wrong could do worse than have a look at the loss statistics produced by the insurance industry. A (more or less) exposed appendage like the rudder is always going to be something of an Achilles' heel and damage to the rudder, its bearings or other parts of the steering system will tend to have serious consequences, so this is a subject the insurers hear plenty about.

A rudder attached to a skeg and robustly mounted with a third bearing will clearly do better in extremis. Even if the lower part of the rudder is lost, enough of the more protected upper part usually survives to provide at least some control.

The Golden Globe Race

It seems perfectly logical to me that traditional designs are back in favour. The entry list for the *2018 Golden Globe Race* (GGR), for example, reads as follows:
- Rustler 36 (6)
- Biscay 36 (4)
- Lello 34 (3)
- Belliure Endurance 35 (3)
- Suhaili ERIC design replica (2)
- Ta Shing Baba 35 (2)
- Tradewind 35 (2)
- Nicholson 32 Mk X (1)
- OE 32 (1)
- Benello Gaia 36 (1)

Why has what was once regarded as ideal now fallen out of favour in many quarters? Is it simply that boatbuilding has changed and what was once a craft is now an industry – an industry that has no qualms about jettisoning time-honoured principles if it boosts the bottom line and that has a willing accomplice in the form of revenue-hungry glossy magazines to help it spread its message that the new generally beats the old? Surely not!

I'm not one for wallowing in nostalgia, I assure you, and I'm positively enthusiastic about innovation, but when it comes to sailing I draw a very clear line between *a bit of fun* in friendly waters and *more serious undertakings* far from safe harbour. I believe this line has become blurred in the head of far too many sailors today, partly because it suits them to think that way, perhaps, but also because they lack the knowledge (or good advice) to appreciate the differences.

Dramatic as it may sound, set off on a long trip with the wrong type of boat and whatever else happens, you will need a measure of luck to come home safe. Set off in the right type of boat, on the other hand, and you need only worry about your own seamanship and the practices of the prudent mariner. Keel and rudder are always the first point to check. A quick glance around the deeper recesses of your own brain should be enough to start you off in the right direction …

Treacherous waters

My belief that national preferences regarding designs, construction and builders are largely the product of media reporting, targeted marketing through advertisements, orchestrated test reports and *boat shows* and the observations and recommendations of well-known sailors is thoroughly borne out by the clear differences that exist between countries and continents.

Robustly built traditional types with a long keel and protected rudder are far, far more popular in the USA, for example, than they are here in Europe. I suspect that this is due to the large number of successful American blue-water sailors who have stuck with the traditional model (despite the fashion for "progress" and intensive marketing campaigns promoting a very different type of boat for bluewater use), have found it very effective and have written of their experiences in books that their fellow sailors find credible and compelling.

I have nothing against compromise, at least when compromise becomes unavoidable, but I would have a very limited appetite for compromise in a matter as grave as choosing the right boat for bluewater adventuring. A solid rudder robustly attached to the boat would seem to me an elementary requirement.

Am I over-egging the pudding somewhat here? Perhaps I am, but if my efforts are complicating your decision-making process or causing you to revisit past decisions, so be it: ordinary life is all about compromise, but there are different rules at sea and they are not of our making. That's my view and if you were to judge me stubborn for it, you wouldn't be the first!

Emergency tiller

A large proportion of the boats on long-distance cruising duty rely on a wheel steering system. Presumably few would deny that the transmission components making up such systems constitute a weak point: after all any failure here can have very serious consequences. This leads me to view wheel steering systems – for all that they represent the state of the art for most designs – as another of those compromises I mentioned. Begging your pardon, I have to say wheel steering does rank in my mind as something of a *minor sin*, at least if the boat concerned lacks an effective emergency tiller. Remember the helm is not normally staffed on long trips anyway thanks to that little marvel of engineering that spares bluewater voyagers the purgatory of endless manual steering.

It used to be standard procedure for the main rudder shaft to extend all the way up to deck level so that a good, solid emergency tiller could be fitted quickly if necessary to restore proper steering. This had the added bonus in many cases that the windvane self-steering system could be connected to the emergency tiller rather than the wheel to keep the transmission lines short and minimise friction and inconvenience (a win-win if ever there was one). It is still quite common in France to see boats on which the wheel is connected directly to the tiller by lines routed on deck that can be removed (via shackles) when at sea – because *humans don't helm at sea* in any case.

The industrialisation of large-scale boatbuilding seems to have somehow changed the status of the emergency tiller to the extent that one risks being perceived as a bit of a wet blanket for even suggesting it ought to be robust and effective. I don't wish to trouble anyone unduly, but in my opinion the standard emergency tillers offered for many mass-produced yachts are just not up to the job of controlling a boat in heavy seas.

I consider these half-hearted solutions a *sin* on the part of a boatbuilding industry that presumably manages to get away with it simply because sailors are no longer sufficiently aware of the importance of still being able to control their heading if the wheel-steering mechanism is disabled. Astonishingly this matter is barely even discussed in the German-speaking parts of the world despite seeming to feature fairly regularly in English-language publications. Food for thought!

Drive systems

From the harbour tug operator to the car driver with an empty tank, anyone with first-hand experience knows the best way to set a large load moving in the right direction is to push it from directly behind or pull it from directly in front. Ship propulsion systems have followed this principle ever since the engine below deck first took the place of canvas and oars and made shipping rather less dependent on the weather.

A propeller pushes in a straight line against the shaft and the ship's engine at the end of it to move the whole show forwards while wasting as little as possible of the force generated. The force applied is obviously substantial – and obviously this force really ought not to be transmitted through the engine mounts, as they are not designed to handle shear forces. Just imagine what would happen on the average sailing boat if those relatively insignificant little rubber and metal feet under the engine had to bear the brunt of every powered manoeuvre! Now imagine what would happen if the boat was unexpectedly brought to a halt due to grounding or a collision, for example. Suddenly all of that force driving the boat forwards through the water would instead be driving the engine forwards through the saloon. And nobody wants to see that ...

The thrust bearing was invented to solve just this problem: robust as can be and permanently connected to the hull, thrust bearings efficiently transfer the thrust produced by the propeller to the boat while leaving the engine to work away at its shaft in peace untroubled by any thrust forces and safe and sound on its mountings. It's just straightforward, logical physics.

There was a time powerboats and sailing boats alike played by these rules, but then one day the bean counters decided to revisit the manufacturing process and resourceful boatyard owners caught wind of a new way to boost their earnings. Yacht propulsion systems are a thoroughly complex affair involving a large number of components:

- Engine – transmission – selector gate
- Shaft drive plus thrust bearing
- Stern tube with or without lubrication or seal

- Tail shaft, propeller and perhaps a cutter too to mince and expel stray ropes and nets without disturbing the skipper's peace.

Obviously installing something this complicated is expensive and time-consuming – and hence an attractive target for cost-cutters and efficiency freaks. People are seldom more inventive than when finding ways to tilt the cost/income balance more firmly in their favour!

Shaft drives

Any displacement of the thrust axis relative to the angle of the crank shaft means a loss of efficiency, because redirecting the force from the engine (transmission using bevel gears) adds friction that reduces the drive reaching the propeller. It also adds torsion to the system: the thrust forces are opposed by shear forces that can no longer be directed straight into the hull because of the leverage created by the mismatch between the thrust axis and the angle of the crank shaft. This all happens safely out of sight on a yacht, but outboard motor users will probably be quite familiar with the way the prop shifts forwards as soon as forward gear is engaged. The thrust forces generated at the bottom of the shaft are transmitted to the hull at the top via the transom. Every change in thrust, forwards or backwards, causes the position of the prop to change. This action can be vigorous enough to cause serious problems for a weak transom.

If you were ever (quite by accident of course) to give the outboard on a Banana boat or similar folding dinghy a few too many revs (the safe limit is around 2 HP), you would see how the whole boat bends and twists, rocking over every wave as the motor pushes in one direction and the hull tries to recover its shape in the other ...

Z-drives and saildrives

Z-drives and *sail-drives* were a logical development for companies anxious to slash the cost of installing and configuring inboard engines. Once the hole in the hull has been prepared and the pre-assembled engine and drive components have been inserted and bonded into place, final system installation is a remarkably rapid matter of simply connecting up the various pre-assembled modules.

This is not the place to go into the details of Z-drives for powerboats. Suffice it to say that the solid back end typical of this type of boat readily accommodates drive components and transmits the force from the propeller very effectively into the hull, especially as the Z-axis can generally be kept quite short (after all powerboats almost always travel with the bow high and the stern low).

Saildrives

Hardly any high-volume yards still use conventional shaft drives: the cost advantages of saildrive units are just too great for builders to ignore.

The owner whose sailing is limited to weekends and holidays and whose boat spends a reasonable amount of time on land every year has ample opportunity for service and maintenance activities and therefore a different set of priorities to the sailor who expects to spend prolonged periods afloat with no access below the waterline.

The only component of a conventional shaft drive system exposed to the sea is the propeller, but sail-drives have sensitive components – the entire Z-drive – located underwater out of reach. These components need servicing and oil needs changing but it is generally impossible to do this or attend to any leaks or damage without hauling the boat out, which can be a far from simple proposition in many parts of the world.

I once had word from Bobby Schenk, who was sailing off Malaysia at the time, that his *Privilege catamaran* was suffering from "mayonnaise" (caused by seawater mixing with the oil) in the saildrive transmission. A cunning solution supplied by knowledgeable fellow sailors eventually saved the day.

The underwater parts of saildrive units are almost always made from aluminium, a material that can have an unhappy time of it in seawater unless the *AlMg5* alloy is used. Zinc anodes are therefore essential to protect the drive components, but these anodes require regular replacement, which creates a further problem for users doing their sailing far from the nearest big boatyard. Saildrives need regular servicing to perform reliably too and that means regular haul-outs, but since lifting the boat out of the water isn't an option in many parts of the world, this has to be regarded as another potential Achilles' heel, especially for bluewater use.

The hard facts

I have done my best to produce a balanced assessment and searched high and low for advantages to using a saildrive on bluewater trips, but I just can't unearth any. Conventional shaft drives can provide good manoeuvrability too so long as the prop is sited close to the main rudder. The list of saildrive disadvantages, on the other hand, goes on and on:

- You need a big hole in the bottom of the boat and have to rely on a rubber seal (or two) to keep the sea out. A traditional stern gland does allow water ingress, but only a drop or so every minute. A failure of the saildrive seal, on the other hand, is a nightmare

scenario and manufacturers and insurers alike insist on a strict inspection and replacement regime.
- A saildrive is much more exposed than a conventional shaft drive to whatever happens to come drifting along in the sea (submerged rubbish, ropes, nets) and to damage in the event of collision or grounding.

Volvo's IPS – POD drive systems for powerboats are designed so that the entire underwater part of the system will break away from the hull in the event of a collision to protect the watertight integrity of the hull (better a boat with no engine than no boat at all). Saildrive units usually have a longer lever than pod drives, so there are more options for a potential *designated failure point*. Not that any of these options is particularly enticing:
  o Shear/break at the flange below the seal
  o Overloading of the saildrive mount in the bottom of the boat
  o Break in/distortion of the bottom of the hull

WHY "ACHILLES' HEELS"?

When choosing a boat for extended voyaging, as little as possible should be left to chance: trusting to luck is never going to be as sound a strategy as removing the potential for problems in the first place. The term *Achilles' heel* accurately conveys the seriousness of the issues under discussion.

Things that go bump in the night

Knowledge, information, what the military likes to call "intelligence" – finding out us much as we can that's relevant to our plans is usually one of the cornerstones of sound planning, but as sailors there are some facts and figures we prefer to overlook. The number of shipping containers lost at sea around the world, for example, runs into the thousands even according to the official data (which seems likely to underestimate the scale of the problem significantly).

Should we regard bluewater sailing as somewhat akin to roulette then, or do we think these random steel reefs probably sink to the depths quickly enough for us not to worry? How many of them settle just below the surface, lurking out of sight but well in range of keel and rudder until the product or packaging that prevents them sinking finally becomes waterlogged or breaks down?

At the risk of leaving myself open to the charge of being a conservative old duffer opposed to any and all innovation in yacht design (for example), I am going to stick my neck out here and draw a clear line between sailing as *fun* and sailing as *serious business*. Part of the difference has its roots in the distinction between *racing* and *cruising*. Prioritising absolute speed and records over anything else because that is only way to persuade the *CEOs* of major sponsors to put their hand in their pocket and hold the attention of armchair fans not already *wedded to the sport* leads to decisions that would be difficult to rationalise in other circumstances.

The situation in the world of cruising looks rather different, because boatbuilders design and cost their creations for optimal mass production and bluewater sailors with farflung destinations in mind are inevitably not the main target market. Unique selling points have to cover as many different potential user types as possible. Space, comfort and, most important of all, performance have to take priority, especially in light of the fact that the potential customer often has less enthusiastic associates to win over as well if the sale is to be completed. It all comes down to compromise, not surprisingly, but knowing when to compromise and when to stand firm can be difficult without good, honest, unembellished advice.

It troubles me that the "modern" designs dominating today's boat shows and thronging our marinas (*performance cruisers*, the marketeers like to call them) are being recommended to – and used by – sailors with long-distance ambitions. What about the keels that fall off, the rudders that fail, the boats that simply disappear and the entire fleets that end up flat-packed on some otherwise peaceful shore after hurricane season? Natural selection? Economic stimulus programme funded by the insurance industry? Sailing as a game of chance? It all depends on one's point of view. Or should that be the firmness of the ground beneath one's feet? But then again ...

Murphy legislates

The *KISS* principle can always be relied on because everyone knows the complicated things always go belly up first. It makes plain good sense to find the right comprises between (relative) sophistication/luxury and reliable simplicity – compromises that sit well with everyone involved (should anyone else be interested) and not just the prime mover.

Think about speed, for example: a *faster* boat might get you there 24 hours earlier, but what if the price to be paid for those 24 hours is a bumpier ride at sea, more noise down below and consequently a sleep deficit to make up once in harbour? Would the need for a night or two's undisturbed sleep in a hotel at the end change the balance at all?

I am convinced that people make better decisions if compromises are acknowledged and described as such from the outset (which is why I'm writing this). I understand that tackling everything head on in this way doesn't suit everyone, but a compromise is a compromise whether we admit it or not, so we may as well continue.

Soft impacts

The sea abounds with objects that would like nothing better than to envelope a nice saildrive, snuggle in between skeg and rudder or wrap themselves around a water generator or submerged parts of a windvane steering system, for example. Plastic bags, fishing nets and old ropes are not the only culprits either. Even seagrass can make its presence felt, especially when encountered in large quantities. How heart-warming it is to think that when a long-keeler meets a mystery something in the water, it just pushes it down deeper below the surface, more or less gently, until the boat has passed and the grass, if grass it was, can gently rise back to the surface and the sunlight and get back to growing again. Without *hitching a ride* on any appendages.

Long-keelers are what might be called self-cleaning in this respect, whereas modern designs tend to snag anything in the sea that can be snagged. And the crew of a boat with a balanced rudder has no real way to tell how much seagrass their rudder has *harvested* because it is hidden away under the hull.

Engine cooling water

Finally, I ought to mention that plastic bags of all kinds – and jellyfish as well for that matter – can potentially block the intake for engine cooling water. When motoring at low speed, for example while hunting for a good spot to drop anchor, a weak suction effect can cause objects like this to cover the intake port so that no water can enter, quickly leading to overheating. The solution is to stop and restart the engine and then proceed at speed. How easy would it be to build boats with a second cooling water intake on the other side of the hull so that the crew could simply switch from one to the other in such situations without cooking the engine?

The sailor's life is full of exciting moments. Being prepared for them makes it possible to avoid at least some of the negative consequences for boat and crew. Forewarned is forearmed.

TRANSOM ORNAMENTS

Everybody knows that appearance-wise, one sailor's dream yacht is another's blot on the seascape and that we all have our own personal priorities. My professional interest centres on the after end (where else can a Windpilot go?). Like some Don Quixote of the pontoons, I have been impudently annexing the prime spot in the middle of the transom for decades.

Sometimes it takes all my cunning and reason (not to mention a measure of courage on occasion) to bring people accustomed to knowing their stuff and having their way (including the odd know-it-all) round to my way of thinking. It can be a painful process (painful, that is, for people who haven't spent a large part of their life wrestling with the issues involved). I have become used to sharing my opinion in direct fashion. Politeness has its place (so they say), but the risk of being misunderstood or brushed aside is just too high. An opinion merits airing only if one has the wherewithal to back it up and an audience able to digest it. What possible use could my advice be to someone who knows they know everything already?

Antenna arch, antennas, wind generator/hydrogenator/solar panel(s), dinghy, davits, outboard, bathing platform, radar mast, bimini ... oh and one of those windvane steering things: that's the sailor's basic wish list when it starts to get serious and thoughts turn increasingly to distant waters, sunshine and palm-fringed sandy anchorages.

Balancing the natural desire for cold beer, decent food and a bit of shade for delicate Northern skin with essential safety considerations and the need to maintain at least a certain level of creature comforts for any less hardened members of the crew without overly compromising the aesthetics of the pride and joy can be a tall order – often too tall even for the bluewater specialist yards. The job of deciding what to take and where to fit it ultimately falls to the skipper (the person who wields the real power I mean, not necessarily the one who pays the bills). Naturally decisions with significant financial as well as technical consequences often go to committee, at which point considerations other than direct performance at sea (and creative negotiating tactics) may have an impact ...

Whoever ends up with the final decision faces the challenge of ascertaining, ordering and reconciling the great mass of conflicting requirements, requests and interests to produce a vessel suitable for global travel that works flawlessly at sea without intolerably offending the users' taste. Producing a boat that manages to do all of this and still look good even without the proprietor's rose-tinted spectacles represents an achievement worthy of real respect. Get it wrong, either due to a failure to think things through properly or a simple lack of knowledge, and the results can be inelegant to say the least. Done badly even a new build can press all the wrong visual buttons.

Compromises cry out from every angle and yet ... and yet there is another way (another way apart from the financial torture of a *one-off built to order*): it pays to ponder the details and be intellectually rigorous in the search for effective solutions.

I spend a good part of the time devoted to helping sailors (indeed a good part of my life) on marking out the boundary between the worthwhile and the pointless and on ordering and reconciling priorities, mainly to ensure (insofar as it falls within my power) that the boat of the moment does not spark too sharp an intake of breath when a well-meaning but still critical gaze falls upon the hind quarters and the army of afterthoughts there to be found. This area can prove treacherous territory for human hind quarters too, from trousers terminally shredded by the merciless bite of a deck fitting to the tragic demise of a harmless little antenna obliterated under the impact of an ample and unwieldy rear end.

The world's bluewater library offers countless books identifying, describing and underlining the important features of different bits of kit and while this may, perhaps, help readers to avoid errors in their decision-making, there is to my mind a distinct lack of more detailed practical advice and suggestions as to how these different systems are to be installed and operated together in the same small space on a real boat on the real sea.

It must surely be inevitable that leaving the stern of the boat as-built to the uncoordinated attentions of a succession of fitters, each concerned only with installing one specific item of equipment, will seldom result in visual harmony. It seems to me that we lose all objectivity when looking at our own boat (although mine have always been genuinely beautiful) and that not all of us have an interest in learning the error of our ways either.

What would cars look like if we treated them the same way we treat boats? Picture a dashboard crowded with retrofitted instruments randomly screwed, taped or clamped to every bit of available surface. Would anyone tolerate that? When it comes to boats, on the other hand, many people seem quite prepared to forget about looks altogether or to assume – mistakenly I might suggest – that a yacht's effectiveness as a status symbol inevitably *increases with every new gadget on display.*

Rewind 40 years and a *Windpilot* at the stern and *baggywrinkle* in the standing rigging would have been quite enough to mark out a bluewater veteran. Today's prospective long-distance explorers, in contrast, have an almost endless list of accessories available to them. Which is handy in one way: the imponderables of existence prevent many sailors taking the final step to realise their dreams, but so long as the boat is still there sporting its many add-ons with intent, they can continue to insist that departure is merely postponed and not cancelled. "Other things being equal we could leave tomorrow, but we just need to sort out X, Y and Z first. Hopefully it won't be long now …" And all the while, with each delay leading to another round of planning and procurement, the owner gets to bask in the reflected glory of a yacht that is visibly highly equipped for the challenges of the cruel sea to come.

Is there simply not enough demand in the market for smart solutions? Do sailors just not think about this aspect? Or is there a general lack of awareness that – and of how – things could be different?

It baffles me that prospective bluewater sailors are so often left to their own devices in this area and I hope that sharing my own thoughts and suggestions about how best to accommodate the necessary equipment will go some small way to making good the deficit. Encouraging everyone to give a bit more thought to this issue might even make some of my future face-to-face discussions a bit more straightforward too – I live in hope!

Windvane self-steering system

The pendulum rudder on a servo-pendulum system is potentially vulnerable to impacts soft or otherwise – even a large mat of seagrass – if its mounting does not allow it to swing up when overloaded. A fouled pendulum rudder can experience considerable drag, which will reduce self-steering performance significantly and could even lead to failure. Interestingly, the sight of thick clumps of seagrass trailing from their self-steering system is often the first indication the crew has of this particular hazard (and, one would hope, gives them cause to stop and consider the situation of the main rudder and/or saildrive).

Time at the back

Speaking with manufacturers and boatyards that welcome suggestions and are prepared to give consideration to and implement ideas and improvements from outside is always a pleasure (and that pleasure is its own reward: I share my input on these matters as an idealist interested in the good of bluewater sailing and not in pursuit of personal financial gain). A little cooperation can go a long way, as in the case of my work with Armin Horn, inventor and manufacturer of the *SailingGen* hydrogenerator that did sterling work for two of the boats in the *Golden Globe Race*.

Boatbuilders too are always keen to hear the details about possible ways to improve layouts at the stern, although it has to be said the French aluminium specialists are particularly innovative anyway when it comes to ingenious solutions –and they do good business too, coincidently!

Compared to the levels of refinement to be had with French aluminium bluewater yachts, many of the major international boatbuilders are still living in the dark ages. It is surely no accident that so many GRP boats find themselves saddled with unconventional add-ons detrimental to their visual appeal.

Local strength

The very first point to consider when installing accessories is the local strength of the installation surface, which gives metal hulls an instant advantage over composite constructions in terms of the options for finding a solid base for antenna arch, davits and stern rail. The (all too common) sight of insufficiently robustly mounted equipment precariously perched around the place always makes me feel queasy.

Pole-mounted radar sets and wind generators often experience significant vibration and, when bolted to a GRP deck, can work loose or even break away entirely at sea.

A radome riding three metres above deck height can develop considerable inertia on a boat bouncing around in rough seas and it doesn't take a physicist to work out how this could end if the pole is not anchored firmly enough or if the loads involved are insufficiently well distributed. Arrangements in which the pole relies on wire or tube supports that only cover two dimensions allow the development of dynamic forces that can pose a severe test for any mounting.

Far too many stern rails are attached to the deck with undersized fastenings, some are not even through-bolted and some lack the necessary reinforcement/backing in the relevant areas of the deck. Using a rail that is itself inadequately mounted to accommodate equipment such as a life raft, for example, tees up a sequence of events that could prove catastrophic. No less dubious are davit mountings that concentrate too much load in too small an area, especially if the corresponding section of hull/deck is not reinforced for such purposes.

Swim ladder

The first signs of disquiet usually begin to creep across the owner's face with the news that that centrally mounted swim ladder is going to have to move if a windvane self-steering system is to join the party (the few exceptions serving only to prove the rule). No sooner has this mental hurdle been overcome (with my generous assistance) than a second shock arrives: a ladder is never used in anything of a sea in any case and especially not for recovering an MOB, as the bouncing stern of a boat in waves offers no safe refuge for a person in the water and legs and feet need to be kept well away from the prop. An MOB should always, always, be recovered over the side. There are all kinds of solutions available including the pilot ladders that have been mandatory for our French neighbours for years, for example. Mention such things in Germany and the response tends to be a blank look.

Once we accept that swim ladders are not safety-critical, the all-important question of how to combine one with a windvane steering system is easily and elegantly resolved: swim ladders are almost always attached with just two bolts, so a few moments with a wrench is all it takes to remove the ladder ready for passage-making, where the vane gear is king, and put it back again afterwards. And relax ...

The discussion about mounting windvane steering systems off-centre is one I expect to follow me to the grave, but I am happy to dive back in again should anyone feel the need. Just e-mail me.

Gangway

Essential in the Med, where reverse parking counts as state of the art, the gangway (or plank) is more or less obsolete beyond Gibraltar (but don't forget the pontoons in Las Palmas). Generally speaking, the less sophisticated the catwalk, the more widely it can be used. A plain three-metre plank takes some beating in this regard, as it can also be called into action as a fender board should it ever be necessary to tie up alongside a concrete or sheet-piled harbour wall. Monsieur Henri Amel did his admirers – especially his customers (and indeed himself, given that he was blind for a large part of his career) – a great favour by inventing a ladder/gangway integrated into the stern rail that can be used anywhere. This solution can be emulated on other boats too with a little investment (see, for example, Sybille and Christian Uehr's solution for *SV Subeki*.

Bathing platform

An open or fold-out bathing platform is one of the characteristic features of modern yachts. These platforms have all kinds of uses and make a worthwhile addition (permanent or folding) to traditional designs. A retrofitted platform should sit about 40 to 50 cm above the waterline. It is important that the platform extend no closer than 25 cm at most to the edge of the transom, as otherwise it may drag in the water when sailing fast and heeled.

There is much to be said for suspending a platform *from above* rather than supporting it from below on brackets because anything below the platform will be wet all the time and any through-hull fixings in this area could leak. Another advantage is that struts or brackets attached to the hull above the platform provide something against which to prop containers and shopping bags/boxes safely so that they don't fall into the water.

Platforms should be *modular* in design so that they can be repaired/removed/replaced easily even far from home. A *bespoke* one-piece platform might look better but will put a big dent in the budget and could be very difficult to repair on the hoof (not to mention the unnecessary extra weight).

Windvane self-steering systems can coexist happily with bathing platforms of all descriptions. Smart vane gear manufacturers have a solution ready off the shelf (or up their sleeve) for all stern configurations, which is not only very useful up front, but also very handy in the event of damage. I think that's probably all that needs to be said here on this subject.

Davits

The first matter to clarify is that davits, which seem to have become a burning *must-have* for people cruising offshore on anything longer than about 35 feet, have no business dinghy-dangling at sea other than on the largest of yachts: one *unscripted encounter* with an overenthusiastic wave and – hasta la vista dinghy (and davits) – you have a blank slate at the back ready for the next crop of gadgets (and, for the time-being, more space to sit).

A dinghy is a very big target for incoming waves and can consequently transmit enormous forces to the davits and onwards to the deck. If there is a weak link in the system, be it the davits, the way they are fitted or even the dinghy itself, there is a good chance the sea will find it. The inertia associated with a dinghy should not be overlooked either – truly there is ample opportunity here to get it wrong! Davits are ideal for parking the dinghy overnight or transporting a wet tender during short hops from anchorage to anchorage, but at sea they are best regarded as a decorative feature, or at most a useful place to stick solar panels, wind generators or antenna bases.

Dinghy

Choosing the right dinghy can be tricky. The first thing to bear in mind is that while they are certainly convenient to stow, compact folding or inflatable tenders with a collapsible/inflatable floor are only suitable for small outboard motors or rowing. They may be fun in a sheltered marina close to home, but unfamiliar waters and exposed coastlines demand something with a bit more to it. Fighting your way back to the mother ship in deteriorating weather in a flimsy tender powered by just a small outboard or your own efforts at the oars can be an ordeal. Every crest knocks you off course and the advantages of something more robust suddenly become very clear.

A RIB equipped with a small motor is stiff enough to plane even in rough water, making travel under adverse conditions much less of an adventure. RIBs, which also have advantages in terms of safety and lend themselves to towing (at any reasonable kind of boat speed), are the standard solution across the anchorages of the world.

As an aside, it makes sense to buy your tender when you are actually going to need a tender and not before. *Caribe RIBs*, which are manufactured in Venezuela using Hypalon, have a good reputation for robustness and affordability. There is a lively second-hand market within the sailing community too, which is worth considering not least because a pre-loved dinghy is less likely to be stolen. Where theft is an issue, a dinghy that spends the night out of the water – whether in davits or on the main halyard – is much more likely still to be there at breakfast time. What the voyaging yacht owner often regards as no more than a useful accessory can represent a significant catch for a light-fingered local in many parts of the world and the trick of disguising a brand-new tender with an old tarpaulin over the tubes and a splash of paint on the engine cover is universally known – including among those none-too-rare jokers who will politely guard your boat all day only to relieve you of it after dark if they judge your tip to be insufficient.

Outboard motor

Horsepower? Suffice to say a RIB will plane with an 8 HP outboard. Outboards are best stored below deck rather than on the stern rail when not in use, as the aluminium alloy used *tends not to be sea-water resistant* and its paint job is thus the engine's only protection against corrosion. An engine left on display, moreover, can make a tempting target for anyone minded to acquire one for free.

I know that some people object to *more powerful* outboards because of the disturbance their noise and wake cause for other boats anchored nearby, but on the other hand faster-moving dinghies are *gone sooner* – and then there is the added safety factor that comes with a more powerful outboard to consider too, especially when you are the one driving.

Antennas

A forest of antennas (ten is an entirely realistic number now for a well-equipped vessel) is one of the prices to be paid for 21st century safety and convenience at sea. The preferred mounting locations are
- Top of the mast
- Leading edge of the mast
- At the stern on the backstay
- At the stern on deck
- At the stern on the stern rail
- At the stern on an antenna arch

Required/desired range is the key factor in determining what ends up where (bearing in mind that there will be little space left at the mast head once the wind instruments have been fitted). The four main options are the leading edge of the mast, at the stern on a pole, at the stern on a framework of stainless steel tubes and at the stern on an antenna arch. Height above water is irrelevant for satellite antennas as the signal comes from above.

Wind generator and solar panels

Proper location is essential for wind generators and solar panels: the former need a good airflow and enough space to spin without mincing the crew and the latter need plenty of light. Solar panels can be installed in many different locations or even moved around for best effect.

This is the point at which the idea of an antenna arch should begin to seem really rather attractive even to those initially reluctant. An antenna arch combines practicality and safety, can perform multiple roles and has the potential to look good doing so too.

Antenna arch

Invented in France, where it is known as a *portique*, the magic ingredient needed to satisfy all of a sailor's wishes effectively in one smart package and keep the stern looking good is an antenna arch. Already established as a characteristic feature of *Ovni* yachts 20 years ago, the antenna arch quickly became a common sight among French bluewater sailors despite its unconventional appearance.

I consider the antenna arch an excellent invention: it makes installing important equipment very much simpler, has significant benefits in terms of safety and is now very smoothly integrated into the deck layout of French yachts.

All new *Allures, Boreal, Garcia* and *Ovni* models come with one already in place.

Advantages:
- The cables linking antennas to the corresponding device at the nav station are kept relatively short
- Antennas will continue to function after a dismasting (unless the mast comes straight down over the stern, in which case the arch at least provides some protection for crew in the cockpit)
- The radome is high enough up to avoid irradiating the crew
- The arch can also be used for other accessories

The logical next step in the evolution of the antenna arch was to integrate other accessories as well. *Garcia*, for example, made the integral davits mount standard on its designs a long time ago and I myself have been speaking for years about the advantages of a *bimini* that is made to match the arch and can be attached to the front of it in such a way that it folds away onto the arch like the cover on a baby buggy when not in use. Folded out in the manner of an umbrella, a sunshade of this type can withstand even quite strong winds (so can be used while sailing) but keeps the cockpit coaming and side decks completely free of stays and guys.

The arches featured on the French aluminium yachts are always very robustly welded to the hull and it is not straightforward to replicate them on GRP boats because the deck or cockpit coaming laminate (sandwich) will not have been designed to handle such high loads. Be very wary of any antenna arch that can be set vibrating by hand, as a structure this sensitive could let you down catastrophically once the combined inertia of the various devices it supports gets to work in an unsettled sea. We would do well to follow the standards established by the French designers and yards here, provided that the boat concerned has sufficient strength in the relevant areas to support the arch proposed. Anything on this scale that begins to flex or vibrate at sea is likely to topple eventually – with implications for equipment and crew alike.

Antenna arches on GRP boats require effective load distribution and additional reinforcement (of the structure itself as well as its mountings) to prevent the forces brought to bear on the deck causing permanent damage. The antenna arch on particularly wide or large boats can advantageously be sited 30 to 40 centimetres in from the toe rail, possibly as an unobtrusive extension of the cockpit coaming, to leave space for easy line handling during manoeuvres and a handy high-side perch with an excellent view to weather.

Bimini

The bluewater dream is a dream of sun and warmth – and the less sun and warmth you have in your daily life, the more you dream of basking in them on your travels. That much makes perfect sense from my vantage point in Hamburg, but it surprises me how many sailors completely fail to anticipate the downside of trying to live and sail in the glare of the midday tropical sun. A survey of the sun protection solutions on the market indicates that – not for the first time – compromise is the name of the game. Even those investing in new builds, for whom the cost of a bimini is presumably just one more drop in the ocean, have little in the way of attractive options. Good advice appears to be in short supply in this area.

Almost all of the biminis marketed as standalone accessories are sold in kit form, packaged for different lengths or widths, and have to be adapted to fit the intended boat. Most of them come with some significant disadvantages above and beyond the aesthetic too:

- Unstable when deployed
- Tubular support structures susceptible to wind damage (not strong enough)
- Access to the cockpit obstructed by stays/guys that run to the rail/cockpit coaming
- Usually unsuitable for use while sailing due to conflicts with mainsheet, boom or sails
- Available designs seldom match the look of the boat to which they are attached

The number of compromises involved is such that many yachts only ever rig their bimini when at anchor. And time and again the most picturesque of yachts have their otherwise glorious lines defaced by unsympathetic sun protection.

I wish yachtsmen and yachtswomen would expect more when it comes to aft-end accessories. There really are good solutions out there, with many having that intrinsic logicality that makes you wonder why on earth you never thought of them before.

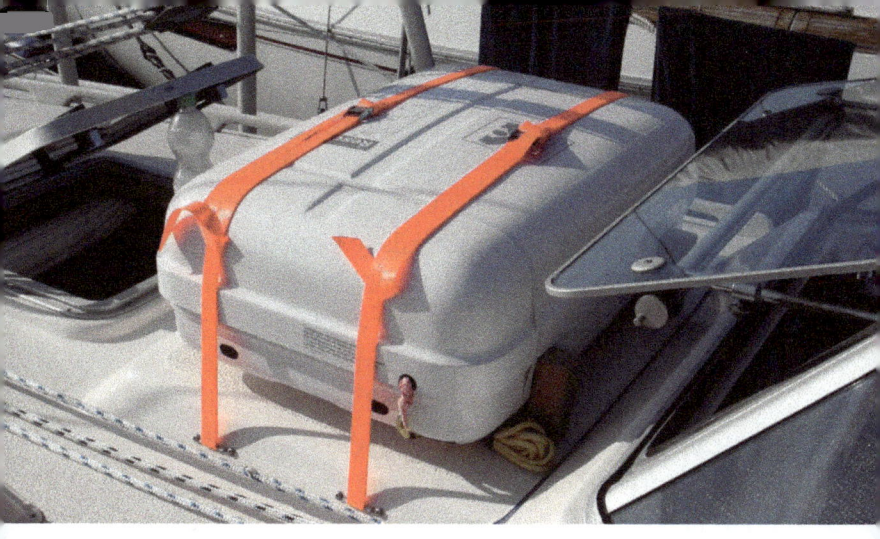

Life raft

Most aspects of the life raft debate have been done to death, but it strikes me we hardly ever hear anything about where best to store the life raft on deck. A large proportion of the life rafts I see have not been installed at all wisely, usually because the installer has given too much weight to the supposed importance of being able to get the raft into the sea easily in an emergency.

Mounting a life raft and cradle on the stern rail might seem an attractive option, but stern rails today are seldom all that firmly seated and the impact of a substantial wave could well overwhelm the self-tapping screws or spindly 5- or 6-mm bolts not infrequently used on modern builds and wash the whole lot overboard, cradle and all. The consequences were the impact to include a crew member caught off balance by the wave as well hardly bear thinking about (which is it say they bear thinking about in considerable depth but only from the safety of a comfortable chair in advance of venturing anywhere far from the coast). The whole idea that the life raft needs to be poised right by the rear exit for easy launching is misconceived: in reality a life raft located anywhere on deck will find its way into the seething brine tout de suite when the time comes.

It is also worth bearing in mind that a container stored vertically will seldom be properly watertight, which is why manufacturers include instructions about making sure the drain holes in the base are opened and kept clear. A raft stored in a permanently damp environment will surely not last as long as one that is reliably kept dry. Whether the need for more frequent replacement is a price worth paying (and whether the manufacturers might have a vested interest in your answer) we must each decide for ourselves.

Life rafts need looking after and they need servicing – and the companies that service them have a very clear understanding of the insurance implications of their work, so if there is any doubt at all about the condition of a life raft, the inspectors will not hesitate to condemn it. Everyone who goes to sea has a strong personal interest in keeping the life raft happy and if that means devoting some time and mental energy to create a better solution – a solution that also helps to save money in the long run – why not?

I like to see life rafts stored flat on deck, preferably under a tarpaulin to protect against spray and UV. French boats often treat the life raft to its own private parking space in the form of a standard recess built into the bathing platform.

Hydrogenerator

Hydrogenerators have traditionally received far less attention than wind generators, but their profile has recently begun to rise due, in the main, to their use on power-hungry fast racing boats and while it may be driven by speed and competition, the market that has developed as a result also has much to offer the more pleasure-oriented bluewater sailor. Hydrogenerators have now become effective enough to render some wind generators functionally obsolete – and this relatively high generating capacity, do not forget, comes from a unit that makes no noise to speak of and represents no danger to the crew (so no more worries about body parts or turbine blades in places they aren't supposed to go).

The full house

The ideal antenna arch has davits in the form of a moveable frame provided on its aft edge and a bimini mounted on its forward edge as well as all the necessary antennas, the radar, a wind generator and movable solar panels positioned to make the most of the sun lined up on the top like so many swallows on a wire. This arrangement keeps heavy cables out of the mast and significantly reduces weight aloft for better handling and performance under sail. It's safer too (see above): losing the mast doesn't mean losing communications. Fewer cables in the mast means fewer cable plugs at the mast step too and fewer cable plugs means fewer of that bane of the sailor's life, dodgy connections. What could be better?

A boat graced by an antenna arch and a bathing platform that
- includes an integral nest for the life raft,
- leaves the centre of the transom clear for the all-important vane gear,
- also incorporates a swim ladder
- and can accommodate a hydrogenerator as well
- will stand out as a paradigm of balance and function in port or at anchor and allow other sailors to see for themselves which errors they might have avoided and what the ideal bluewater yacht looks like today.

I suggest above that low expectations and a general lack of awareness about what can be achieved may well go a long way to explaining the prevalence of poorly conceived and executed arrangements at the blunt end. Reading about it all is one thing but seeing photos of what other people have managed to do and of the smart (in every sense) solutions they have devised should inspire all of us to aim higher and comprise less when it comes ornamenting the inhabited end of the boat.

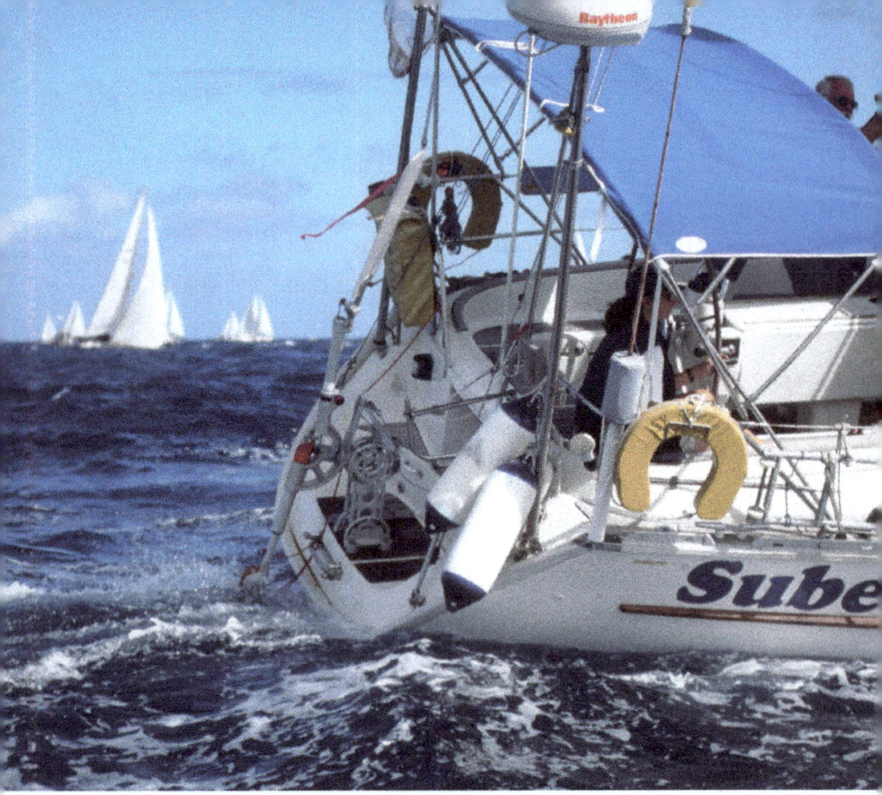
Raytheon
Sube

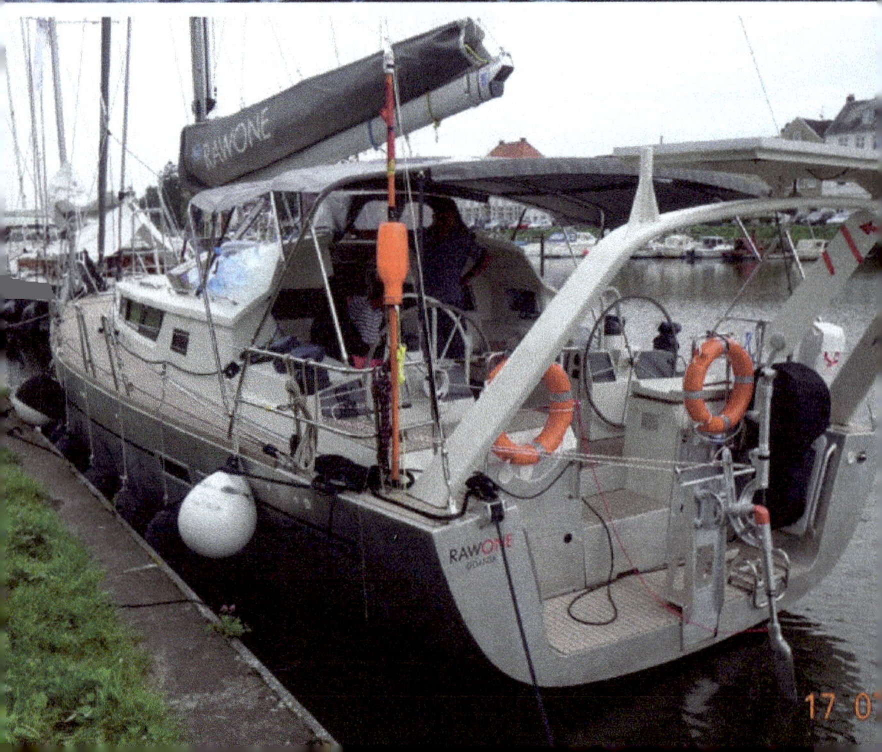
RAWONE
RAWONE

BLUEWATER ADVICE

The 40 years I have spent in the windvane business have shaped the way I think and left me with a clear and obvious set of priorities. A self-confessed idealist, I allow myself the luxury of believing some things still come before sales and profit – and I do so fully cognisant of the fact that this makes me something of an outlier (to say the least). My obvious independence is the *wind in the sails* of the *word of mouth* recommendations that circulate through the cruising community, enabling me to sell my products around the word with no fanfare and no need whatsoever to trouble – much less pay – the usual suspects in the media.

Not surprisingly, certain ladies and gentlemen of the global fourth estate have taken umbrage at my approach because it runs counter to their system of creating dependence and subverts their business plan (salary plus perks), under which paid advertisements tend to correlate with editorial attention.

Mine is a niche business in which the expense of advertising internationally in magazines would have increased the price of my products exponentially without generating sufficient additional interest among sailors. Even before the dawn of the internet, I never felt the benefits would have come anywhere near justifying the cost. The proof is in the pudding: I sell my products worldwide without any conventional advertising and have stayed away from the global *boat show* treadmill since 2003.

Without necessarily having set out to do so, I have taken myself outside the bounds of accepted behaviour in journalistic circles and I believe it still rankles with some that I have no interest in buying *goodwill* to ensure I remain the dog and not the lamp post. Paying for good press has never seemed like good business to me: for all the fine promises of the people whose job it is to sell advertising space, we all know and understand that both they and the journalists they work to support are bound by a very self-evident set of inflexible rules – rules that in my opinion at least are unlikely to leave much scope for putting my interest before theirs.

Not so very long ago I received from French sailing magazine *Voiles et Voiliers* an invitation to book an ad in an upcoming special issue about *sailing two-handed* in which, I was informed, the subject of *windvane self-steering* would command an entire chapter. The pressure to act quickly was on too: ads closed in just a few days! Strangely I don't seem to have received any associated enquiries from the editorial team about my products or company and when I asked about this directly I found no answer forthcoming. As it happens the French market is a big one for me. Hundreds of aluminium yachts from brands like *Ovni, Allures, Garcia, Boreal, Meta* and *Via* have been and continue to be fitted with my systems, some of them even as part of the initial fit-out at the yard. My book on self-steering, *Toutes savoir sur le Pilotage automatique* has been a popular reference source for 20 years and has ensured, together with quite a number of spectacular voyages by well-known sailors, that the Windpilot name is both widely known and highly respected among French sailors. It's the same old story: nothing beats *bouche-à-oreille* (word of mouth).

Thou shalt advertise

The powers that be behind our magazines appear to operate under an unwritten rule that begins and ends with the advertisement. Buy one and you are on the radar and in with a chance of publicity, perhaps even praise. Keep your hand in your pocket, on the other hand, and you simply don't exist. These are the constraints within which our journalists are compelled to operate. Objectivity, it seems, has its price.

Why do I always get the impression that the nautical press is in a constant struggle to justify its own importance? There's just no need for this to my mind. We all (please say it isn't just me) need our regular dose of tales of *heroics afloat* to feed the soul and keep those dreams of gently swaying palms well-watered. The glossy sailing magazines do a fine job of meeting this need by bringing together a wide range of interesting material (including plenty of beautiful, tantalising photos) to nourish our fantasies and supplying this material to us for money.

And magazines certainly know how to do fantasy: sailors get their sailing porn, classic car lovers get their classic car porn, trainspotters get their train porn … and for those with less imagination I expect there is even still regular traditional topic-free porn.

I have gone my own way, something I managed to do comfortably for decades without rubbing the press the wrong way thanks in particular to the fact that there used to be much more understanding of the synergies to be gained from expert knowledge.

I concern myself with counselling the very smallest group of German sailors, namely the bluewater community, whose dreams, plans and experience lend impetus to an entire sector for the simple reason that every sailor, no matter how small their pond, knows what it is to dream the dream.

The very specific questions and concerns regarding equipment and training for the Big Trip are of direct relevance only to a tiny market, but the sailors who make up that market have more than the average share of drive and personal conviction, are hungry and enlightened in equal measure and make their decisions with the meticulousness and prudence that come with realising even one mistake can have wideranging consequences. What this particular group of people need more than anything in their preparatory phase is clear and knowledgeable advice.

Principles

To what extent can specific or even dedicated advice guide the thought and decision-making processes that go on in a sailor's head? Are wise mariners capable of forming their own opinion? Can bad advice actually be damaging? Rhetorical questions all: the answers should be plain to see.

I believe competent advisors, ideally not compromised by the need to stimulate sales, are the best way to communicate complex material.

It seems obvious that increasing monetisation of the bluewater seminars cannot help but have serious consequences for the quality of the advice presented. Allow me to consider the implications in detail.

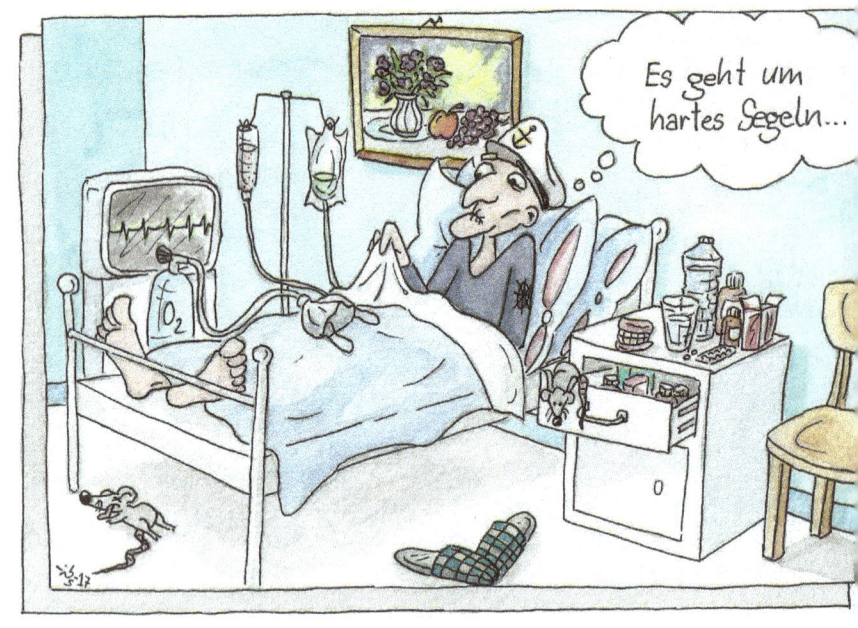

THE PILLARS OF INFORMATION PROCUREMENT

Word of mouth

Word of mouth has become the most important marketing tool bar none for the global sailing community. Opinions, experiences and customer satisfaction are disseminated throughout the community authentically and without artifice – a wonderful reference delivered at the click of a mouse, straight to the target and without costing a penny. But what makes word of mouth the non plus ultra for customers also makes it a long and arduous challenge for manufacturers: earning what cannot be bought takes energy, endurance and sometimes a measure of humility as well. It took me decades of work to bring Windpilot to the position it now enjoys.

Sponsorship

Using *brand ambassadors* to spread the word can seem like an attractive shortcut, which is undoubtedly why it tempts so many. It takes a born gambler though to thrive in a game that demands so much up-front investment when the results are so difficult to foresee and the possibility of the substantial sum wagered disappearing into unhelpful pockets cannot be ruled out.

We all understand how easy it is to create a natural preference for a product by ensuring its name is heard regularly in the right settings. Brand ambassadors for everything from shampoo to the autonomous soapbox have conquered the world; sex, beauty and spectacle have all become a means to a marketing end whose efficacy is beyond doubt. All the world's a billboard.

Sponsoring a sailor is in theory the cheapest way to gain access to the value chain and launch a name out there to be repeated, spread and burnished by the public. Sponsorship can also go spectacularly wrong, as several German companies recently learned to their cost when a high-profile nonstop singlehanded circumnavigation attempt very publicly turned out to be neither nonstop nor singlehanded.

Sponsorship can prove counterproductive in other ways too: people in pursuit of sponsorship have to big-up their project to make it stand out, but if – as can happen – they then end up believing their own hype, they are likely to be disappointed with what any eventual sponsor deigns to offer them. And that does not auger well for the future relationship. Sometimes I suspect each of the parties involved – sponsor, sponsored, media and consumer – has a completely different understanding of how the arrangement really works.

Know-it-alls

All of us who sail started small. Even the most experienced bluewater salts and circumnavigators have gaps in their knowledge that they fill with the help or support of relevant experts (the *shore team*) in order to keep the show on the road. No sailor can afford to pass up the wisdom these experts have to share: quite simply they are a part of a successful voyage.

Nobody ascended from the abyssal plain with absolute mastery of bluewater sailing and certainly nobody will ever be in the position to acquire complete mastery of the subject up here on the surface. Circumnavigators generally view their technical equipment as nothing more than one tool among many: it needs to work for the duration of the trip and that is all. The specialists on whom they rely for their technical support, on the other hand, are usually career experts who have dedicated the best part of their working life to the subject

Now and again a sailor at the end of a circumnavigation will pick writing as his or her next challenge. Reading between the lines of the books that ensue can be an interesting experience: sometimes expert knowledge born of other heads is presented in writing as the author's very own and suddenly – occasionally even overnight – a new fount of all knowledge appears. I consider it to be one of the unwritten rules of respectful human interaction that sources are acknowledged. Giving credit where credit is due does not in any way diminish a sailor's achievement, but it does show appropriate appreciation and respect for the generosity of the people who contributed their technical expertise to the project.

The second question we have to address in our present deliberations is how objective we can expect our sailor-authors to be in print if their experiences have been at least partially enabled by *support*, in cash or in kind, rendered before pen met paper? Can this be accepted, or are those advertising banners the first step in a seamless process to turn advisory services rendered into cash when book meets reader? Once again there is a logical answer to the question.

What does this mean for the book and seminar business? Does it amount to anything more than just one more link in the value chain? The harder I try to put my thoughts in order, the more uncomfortable they become: all of the indicators suggest commercial interests are very much in play and that objectivity could very easily fall by the wayside.

The relationship between author, publisher and manufacturer is undoubtedly a thorny issue on several levels. How do opinions come into being? Who pulls the strings – and who dances when they are pulled?

The media

The capacity to disseminate opinions confers a power to which we are all subject. And the powers that be have a clear interest in exercising it for financial gain. Opinion and vested interests can readily be combined to pave a smooth path to profit, the boundaries between press-pack reporting and marketing dissolving to the point at which they become invisible – or we become blind to them. Barely the tip of a dorsal fin shows above the surface but down there underneath, the jaws are snapping and the tails are thrashing in a bloody fight for influence, money and power – in any order.

The key to spreading a message (or in this case building brand awareness) is *financial investment*, a fact established page-fillers are of course happy to exploit in order to increase their own power – or should that be in order to help justify their existence? The idea of collaboration between the press pack and the book publishing houses is truly a clever one, creating as it does synergies ideally suited to improving the bottom line.

Bluewater seminars

When I joined with Astrid and Wilhelm Greiff (of the German office of the World Cruising Club) in 1997 to trademark the *World Cruising Deutschland* name as a brand for our seminars, we always paid our speakers 500 deutschmarks per presentation. Jimmy Cornell notably declined any sort of payment in order to maintain his independence (as the initiator of such events he had a particular interest in preserving their credibility).

The monetisation of the seminars was one of the consequences of the change of ownership at the top of *World Cruising* at the start of the millennium. In Jimmy's era, it was common practice for years for *event supporters* – of which I was one – to be on hand to assist sailors in port free of charge in exchange for the opportunity to give a presentation to participants in their particular field of expertise. Recent years have seen this arrangement turned on its head: now event supporters are expected to cross the organisers' palm with silver (GBP 5000 it would be for me) for the right to attend to customers in need of assistance.

I had always thought of my inport service visits as an implicit endorsement of the event organisers – an endorsement Jimmy had always clearly valued – and it struck me as particularly perverse that supporters should now be charged to help the very customers who make the event happen. My autumn sojourn speaking face-to-face with real sailors about to take the plunge, a regular fixture from 1976 until 2007, also presented me with a great opportunity to learn about their experiences while sharing my own and I would like to think all sides benefited significantly.

The paywall

Following World Cruising's decision, it quickly became the absolute norm to charge for access to specialist seminars. My refusal opened the door for my fellow windvane provider *Hydrovane*, whose new owner in Canada was then able, on payment of the requisite consideration, to deliver his very special marketing message in person – a marketing message, it has to be said, for a system whose inventor and previous owner, Derek Daniels, had not long before invited me to buy his business (an offer that stemmed in part from his belief that his own product was increasingly unable to compete effectively with mine). My new competitor apparently viewed paying for the right to address his target audience directly as a strategic move that ought to make financial sense.

Choosing my words carefully, it seems fair to conclude that the paywall has certainly benefited the organisers and event supporters while perhaps not serving the interests of the sailors quite so well. Emergency rudders can be a topic worth talking about, for example, but it takes a certain strain of marketing genius to suggest judging the overall utility of a windvane self-steering system on the basis of its purported suitability as an emergency rudder.

Do I regret refusing to pay an event supporter entry fee? Not in the slightest. Not once in decades of visits to Las Palmas and other gathering places of the bluewater crowd in the name of Windpilot have I ever sought to buy or pressurise my way to opportunities. My job has always been to share advice and help sailors get the best out of their windvane no matter who sold it to them.

That I might on occasion find myself taken advantage of by less scrupulous sailors in need of a handyman was a risk I was prepared to accept, but there were limits. One day the lady owner of an *HR 49* on which I was tweaking a vanegear decided to ask me if I could just check the oil in the engine too while I was at it. I wasted no time (or pleasantries) in assisting her to understand that engines were not the reason I was there and by the end of the same day I had decided there would be no more complementary service visits for me. I was 63 years old at the time.

It seems to me in any case that the seminar business has lost its innocence since manufacturers began having to pay for access to a public in search of advice and the right to discuss their products.

Compared with the global cruising community, the German bluewater division remains a charmingly small group of people – which makes it all the more surprising to see jealousies, conflicts and rivalries played out in public again and again. I have the luxury of observing the pitched battles for air superiority and market share from a distance thanks to the fact that Germany has long accounted for only a small part of my own business activities. The perspective afforded by this distance gives me a view I find most troubling.

The advice business, it seems to me, has developed a momentum of its own such that technical expertise increasingly plays second fiddle to vested iterests. Too idealistic? Undoubtedly – but a situation in which the cost of sponsoring equipment can effectively be the *price of entry* to a chain of marketing exposures offers little comfort for those who (with logic on their side) would put competence before marketing budgets.

If mentions regarding technical matters in books and seminar presentations come as the quid pro quo for prior services rendered (in cash or in kind) and if the absence of such a sweetener potentially means important information is omitted, does the concept as a whole have any value at all? Expecting an intellectually rigorous readership or audience to swallow this sort of arrangement strikes me as one faux pas too far. It is perplexing and alarming that anyone was prepared to take this risk.

Sailors tend to be among the sharper knives in the block, a group accustomed to gathering their own information and reaching their own conclusions. I find it disconcerting when unwritten codes of respectful interaction are ignored, especially when the group of people on the receiving end is so small and so discerning. So thoroughly acclimatised have I grown to my own position outside of this whole system that it makes me genuinely sad to be reminded of the mutual backscratching that goes on elsewhere in our field. Time and again, however, there is no overlooking the fact that balance has been lost.

THE DREAM

How would things look if it were up to me? I would like to see bluewater seminars that are inexpensive – or even free – for sailors to attend, that are limited to trusted speakers who know their subject inside out and that are presented by an organiser disinclined to charge its speakers for the privilege of an audience. Such an event would be a big draw for a boat show, for example, as it would be open not just to the elite but to the whole of the sailing community, which would in turn have one more good reason to visit the show in the first place. I would anticipate significant synergies from this approach because it would provide a way to attract new target groups – people who, while not necessarily already planning expansive blue-water adventures, are nevertheless open to the idea and ready to be enthused. Surely it's worth a try?

Actually this approach has been successfully employed for decades in the US, where boat shows offer an extensive programme of seminars. Anyone interested can easily reserve a space for selected presentations for so long as there are still seats available in the venue and the seminar programme ranks as one of the principal draws for many visitors.

Just sour grapes? The niche I have been fortunate enough to carve out brings me fun and financial independence and enables me to enjoy life, so envy is absolutely not my thing. Any notion that the concerns I have outlined above can be dismissed on the basis that it's a perfectly good system that happens not to benefit my business is the reddest of herrings. I just say what I see.

Sailing: fun or serious business?

So where does it all lead? What sort of boats should we be buying to avoid embarrassment at sea as well as in the marina? Which graceful swans – and ugly ducklings – should we avoid if we want to feel confident on the ocean? Will tomorrow still look much like today or will we all soon be gliding over the water on ever more Daliesque foils? Who can guide us onto the path to long-term satisfaction when this morning already seems like yesterday? Are there still reliable sources out there to be heard or have they been lost at sea in the crossfire of talking heads, figureheads, filter heads and influencers?

I am confused. Did I miss an update or a blow to my silvery-grey-haired head? Whatever the reason, reading the sailing media on the latest new ideas and the bleeding-edge techniques and technologies already deemed indispensable for any self-respecting modern sailor leaves me feeling positively queasy. Putting things into context seems to help. I like the metaphor of the iceberg: those sailors who have already cast off are the visible part, those who are still ashore dreaming the far larger part below the surface. For the latter group, everything still revolves around identifying the right boat for the job. Obviously there is a temptation to try to influence sailors and readers a little in their choices, to look at the pros and cons and address some of the issues (without attempting to make things too black and white, of course).

I understand that I cannot cover all areas in a book like this because I would quickly get bogged down, lose my train of thought and, quite likely, my credibility as well. I therefore restrict myself to that small segment of the sport of sailing in which boats set sail with the express aim of disappearing over the horizon and to reappearing some time later, their crew revitalised as never before, in some distant port or anchorage. My interest lies in the craft suitable for such undertakings far from land. Inshore sailors may have different priorities (after all if they run into a trouble – a shortage of beer, for example – the emergency services are only a phone call away).

It troubles me to see how proven principles have been discarded and how a sailing public hungry for information is repeatedly warned in all seriousness that the old rules no longer hold water. Time and again we hear and read damning criticism of boats with a long keel and the claim that they can *trip over their keel* or have no business in the *Southern Ocean* – and time and again these accusations are levelled without the slightest mention of good seamanship, which should be the first and last resort of all who go afloat.

If we are ready to accept that a robust keel with the rudder mounted either on its trailing edge or (with three bearings) on a skeg close behind it no longer contributes to safety and comfort at sea whatever the weather and that a prop on a saildrive is just as simple and reliable as one on a traditional fixed shaft, we are ready to believe pretty much anything the glossy brochures tell us. Believing it though will not make it true no matter how hard we try – and the reality of life at sea tends to lay bare our misunderstandings without mercy.

Generalisations have little value, especially if more detailed consideration is not only omitted but apparently unwanted. Are all of the major production yachts of our time really *suitable* for circumnavigating, but for a few add-ons and upgrades? What magic powers do they have then, if they are some-how to be considered immune to the problems that have halted the progress of similar designs in the past and brought disappointment or even disaster for their crew? The life afloat would be easier if land were the only obstacle to worry about, but as it is, we have shipping containers, logs, dozing whales and plenty else to consider. And that was true even before the orcas added rudders to their hit list.

The contrast between traditional and modern designs in the context of bluewater use is perfectly illustrated by the story of the Woehl family from Flensburg in Germany. The Woehls began with a *Swan*, collecting silverware at regattas and covering some 20,000 nautical miles over a number of extended voyages between 1992 and 2003. The boat lived up to every expectation in terms of both sailing performance and seakeeping. Fate though had a nasty surprise in store for this particular Swan though: sailing along on starboard tack out on the Flensburg Firth one day it was rammed with gusto by another boat. It proved solid enough not to sink, but the damage was severe and the Woehls found themselves boat-hunting once again.

This time they chose a brand new *Beneteau First 40.7* and set about preparing for their planned circumnavigation with a modern *performance cruiser*. The new boat came up short in several departments: the constant abrupt speed changes as it jumped between displacement mode and surfing mode created problems for the wind-vane self-steering system and the boat's motion in a seaway made life difficult both on and below deck. Patience wore thin and an ultimatum ensued: either we get a different boat or we forget about the circumnavigation. The Caribbean is no place to sell a new boat if you want a good price, so the decision cannot have been taken lightly, but soon enough the way was clear for the next chapter in the quest for a suitable replacement.

Salvation came in the form of *Thule*, a yacht that not only ticked all the boxes for bluewater sailing but had actually been built at *A&R* in 1971 while Rainer Woehl was working there as an apprentice. A sensational *one-off* in aluminium and the embodiment of the perfect bluewater yacht, *Thule* still sparkles despite her 50 years. The whole story featured in Germany's *Yacht* magazine, issue 2/2013. *Thule* delivered her crew home hale and hearty at the end of the circumnavigation in 2012 with no damage and no problems.

IN CONCLUSION…

That sailing ventures not infrequently come to a premature end due to *bad luck* is something seafarers fatalistically accept. What we need to ask ourselves though is whether the sequence of events we ascribe to luck might have had different consequences – or even been quite inconsequential – had we done a better job of separating the wheat from the chaff when we had the time. Sometimes probably not, sometimes definitely so.

In an age in which speed records dominate the sailing news and industry insiders and journalists are seriously fantasising about even cruising yachts hopping up on foils and blasting across the waves, we would do well to remember that what goes up must come down, that faster sailing means harder impacts (with UFOs, other boats, land or just the face of the swell in front), that being able to cross it more rapidly does not make the ocean any more hospitable and that speed isn't necessarily the best medicine in any case for sailors who like the chance to sleep and hope to arrive with their boat still in one piece and fit to do it all again.

The fundamental rules of physics have remained unchanged throughout my nearly 50 years of staunchly making the case for better, safer travels at sea. Sailors are no fools, I believe, and facts will always be the best weapon with which to persuade them. There may be no one right answer for everyone, but there are certainly answers that are wrong for everyone (for example in the matter of choosing a suitable boat for extended ocean voyages) and any time dedicated to helping sailors steer clear of them has to be time well spent.

If this book helps in any measure to guide readers through the fairway safe from the shoal waters of misunderstanding and the reefs of misjudgement, that will be good enough for me.

Peter Foerthmann

Lightning Source UK Ltd.
Milton Keynes UK
UKHW020742210121
377372UK00001B/66